# Body Structures and Functions

## Sixth Edition

Elizabeth Fong, B.S., M.S.
Elvira B. Ferris, B.S., M.S.
Esther G. Skelley, R.N., M.S.

DELMAR PUBLISHERS INC.

Cover illustration by Joel Ito

For information, address Delmar Publishers Inc.,
2 Computer Drive West, Box 15-015,
Albany, New York 12212

Copyright © 1984 by Delmar Publishers Inc.

All rights reserved. Certain portions of this work copyright © 1954, 1964, 1967, 1973, 1979. No part of this work covered by the copyright hereon may be reproduced or used in any form or by any means – graphic, electronic, or mechanical, including photocopying, recording, taping, or information storage and retrieval systems – without written permission of the publisher.

Printed in the United States of America
Published simultaneously in Canada by
Nelson Canada, a division of
International Thomson Limited

10 9 8 7 6 5

**Library of Congress Cataloging in Publication Data**

Fong, Elizabeth, 1947–
   Body structures and functions.

   Includes index.
   1. Human physiology.   2. Anatomy, Human.   3. Nursing.
I. Ferris, Elvira Binello   II. Skelley, Esther G.
III. Title.   [DNLM: 1. Anatomy–Education.
2. Physiology–Education.   QT 18 S627b]
QP34.5.F48   1984   612   83-71712
ISBN 0-8273-2185-6

# Contents

Preface/*v*
List of Color Plates and Tables/*viii*

**SECTION 1**
**THE BODY AS A WHOLE**
Unit 1  Introduction to the Structural Units/*2*
Unit 2  Cells/*9*
Unit 3  Tissues/*27*
Unit 4  Organs and Systems/*35*
Self-Evaluation/*39*

**SECTION 2**
**THE BODY FRAMEWORK**
Unit 5  Introduction to the Skeletal System/*44*
Unit 6  Structure and Formation of Bone/*51*
Unit 7  Parts of the Skeleton/*56*
Unit 8  Representative Disorders of the Bones and Joints/*68*
Self-Evaluation/*73*

**SECTION 3**
**BODY MOVEMENT**
Unit 9  Introduction to the Muscular System/*76*
Unit 10 Attachment of Muscles/*81*
Unit 11 Principal Skeletal Muscles/*86*
Unit 12 Representative Musculoskeletal Disorders/*90*
Self-Evaluation/*92*

**SECTION 4**
**TRANSPORT OF FOOD AND OXYGEN**
Unit 13 Introduction to the Circulatory System/*94*
Unit 14 The Heart/*98*
Unit 15 Function and Path of General Circulation/*104*
Unit 16 Pulmonary Circulation/*108*
Unit 17 Blood Vessels/*111*
Unit 18 The Blood/*116*
Unit 19 The Lymphatic System/*127*
Unit 20 Representative Disorders of the Circulatory System/*132*
Self-Evaluation/*137*

## CONTENTS

**Section 5**
**BREATHING PROCESSES**
- Unit 21 Introduction to the Respiratory System/*140*
- Unit 22 Respiratory Organs and Structures/*143*
- Unit 23 Mechanics of Breathing/*150*
- Unit 24 Representative Disorders of the Respiratory System/*156*
- Self-Evaluation/*160*

**Section 6**
**DIGESTION OF FOOD**
- Unit 25 Introduction to the Digestive System/*164*
- Unit 26 Digestion in the Stomach/*173*
- Unit 27 Digestion in the Small Intestine/*180*
- Unit 28 The Large Intestine/*183*
- Unit 29 Representative Disorders of the Digestive System/*188*
- Self-Evaluation/*192*

**Section 7**
**ELIMINATION OF WASTE MATERIALS**
- Unit 30 Introduction to the Excretory System/*196*
- Unit 31 Urinary System/*198*
- Unit 32 The Skin/*205*
- Unit 33 Representative Disorders of the Excretory System/*211*
- Self-Evaluation/*216*

**Section 8**
**HUMAN REPRODUCTION**
- Unit 34 Introduction to the Reproductive System/*220*
- Unit 35 The Organs of Reproduction/*225*
- Unit 36 Representative Disorders of the Reproductive System/*233*
- Self-Evaluation/*236*

**Section 9**
**REGULATORS OF BODY FUNCTIONS**
- Unit 37 Introduction to the Endocrine System/*238*
- Unit 38 The Pituitary Gland/*241*
- Unit 39 The Thyroid and Parathyroid Glands/*245*
- Unit 40 The Adrenal Glands and Gonads/*250*
- Unit 41 The Pancreas/*253*
- Unit 42 Representative Disorders of the Endocrine System/*256*
- Self-Evaluation/*262*

**Section 10**
**COORDINATION OF BODY FUNCTIONS**
- Unit 43 Introduction to the Nervous System/*264*
- Unit 44 The Central Nervous System: Brain and Spinal Cord/*267*
- Unit 45 The Peripheral and Autonomic Nervous Systems/*275*
- Unit 46 Special Sense Organs: The Eye and Ear/*281*
- Unit 47 Representative Disorders of the Nervous System/*287*
- Self-Evaluation/*292*

Glossary/*295*

Index/*307*

# Preface

*Body Structures and Functions,* now in its sixth edition, is a basic textbook which is specifically geared for students of practical nursing and other allied health careers. The text presents an overview of each body system and explains how these systems work together to achieve a balanced state — homeostasis. Basic concepts are developed and amplified through the use of many highly illustrated examples. Clear and concise discussions of representative medical disorders introduce the student to the effects of dysfunction.

This new edition has been thoroughly and extensively updated and revised. The writing style is crisp and unique, making the subject matter interesting, understandable, and easily readable.

## SPECIAL FEATURES OF THIS EDITION

All 47 units of *Body Structures and Functions* have been revised and updated; 18 units were totally rewritten and expanded. The sixth edition includes the following new features:

- a new 16-page insert with 20 full-color anatomical illustrations
- many new and detailed two-color illustrations depicting the anatomical structures that comprise the human body, adding more visual appeal
- more detailed and comprehensive discussions of, among other topics, anatomical terminology, cellular structure and function, physiology, and organ/system disorders
- a greatly expanded treatment of the major body systems including the skeletal, muscular, circulatory, nervous, integumentary, respiratory, digestive, and excretory systems
- a list of key words at the beginning of most units to alert students to their use within the textual material

- a comprehensive glossary of more than 450 words, complete with a new pronunciation guide
- several new tables and charts that help organize complex technical concepts and terms for greater comprehension
- a number of transparency masters (in the Instructor's Guide) which can be used for the preparation of overhead audiovisual materials or handouts

## FEATURES RETAINED FROM PREVIOUS EDITIONS

In addition to the changes described above, the sixth edition of *Body Structures and Functions* retains the following features:

- Specific and measurable learning objectives precede each unit.
- Topics for further study and discussion appear at the end of each unit to promote student involvement in the learning process and to provide opportunities for further enrichment and application of knowledge.
- Where appropriate, guidelines for laboratory study are presented.
- More than 450 unit review questions and assignments promote and test student comprehension and retention of important material and help to identify areas in which students may require further study.
- Self-evaluation tests at the end of each section reinforce and summarize the material presented on each body system.
- An Instructor's Guide includes answers to all assignments and self-evaluation questions, as well as a reference list.
- A separate slides packet contains thirty slides of anatomical illustrations designed to facilitate and reinforce classroom instruction. Two to three slides provide coverage of each body system.

## ABOUT THE AUTHORS

Previous editions of *Body Structures and Functions* were written by the late Esther G. Skelley and Elvira Ferris. Both authors were actively involved in education and had several books published. Esther Skelley was the author of *Medications and Mathematics for the Nurse;* and Elvira Ferris wrote *Microbiology for Health Careers,* with Elizabeth Fong as her coauthor for the most recent edition.

Elizabeth Fong as assistant chairperson of the Biology Department at Brooklyn Technical High School, New York, pioneered a new and innovative biomedical program. In addition to *Body Structures and Functions,* she is also coauthor of the successful second edition of *Microbiology for Health Careers.*

## REVIEWERS FOR THE SIXTH EDITION

Instructors who offered their critical insights during the development of this edition include:

Aileen L. Rowand, M.Ed., B.S.N., R.N., Director,
  Decatur School of Practical Nursing, Decatur, Illinois

Peggy P. Moore, R.N., Health Occupations Instructor,
  Washington-Wilkes Comprehensive High School, Washington, Georgia

Betty Gordon, M.S.N., B.S.N.Ed., R.N., Nursing Instructor,
  El Centro College, Dallas, Texas

Lynn Freeland, R.N., Instructor, Health Occupations
  Lyman High School, Longwood, Florida

## ACKNOWLEDGMENTS

Appreciation and thanks are extended to the following for their assistance and contributions in the development of this edition.

Staff at Delmar Publishers
  Health Occupations Division
    Administrative Editor: Adele Morse O'Connell
    Production Editor: Angela R. Emmi
    Editorial Assistant: Diana J. Miller
  Graphics Department
Illustrators: Joel Ito, Patricia Johnson, Nelva Richardson

In addition, special thanks is forwarded to A.D. Rosenberg, Elaine Fong, and Leon Yankwich.

# Color Plates and Tables

*Following page 120*

| | | | | | |
|---|---|---|---|---|---|
| Plate | 1 | The Ventral Cavity | Plate | 10 | The Systemic, Pulmonary, Renal, and Portal Blood Circuits |
| Plate | 2 | Top and Side Views of the Infant Skull | Plate | 11 | Different Types of Blood Vessels and Their Cross-sectional Views |
| Plate | 3 | Side View of the Adult Skull | Plate | 12 | Blood Cells and Platelets |
| Plate | 4 | Front View of the Adult Skull | Plate | 13 | Respiratory Organs and Structures |
| Plate | 5 | The Heart and Its Valves | Plate | 14 | Alimentary Canal and Accessory Organs |
| Plate | 6 | Conductive Pathway of an Electrical Impulse in a Heart Contraction | Plate | 15A | The Urinary System |
| Plate | 7A | Arterial Distribution | Plate | 15B | Cross Section of the Kidney |
| Plate | 7B | Venous Distribution | Plate | 16 | Uterus, Tubes, and Ovaries |
| Plate | 8 | Front View of the Heart | Plate | 17 | Cross Section of the Brain |
| Plate | 9 | Blood Flow into, around, and out of the Heart | Plate | 18 | Internal View of the Eye |

| | | |
|---|---|---|
| Table | 1-1 | Review of the Life Functions/*3* |
| Table | 2-1 | Units of Length in the Metric System/*9* |
| Table | 3-1 | Different Kinds of Human Tissue/*29* |
| Table | 4-1 | The Nine Body Systems/*37* |
| Table | 11-1 | Location and Function of Principal Skeletal Muscles/*86* |
| Table | 17-1 | Principal Arteries/*112* |
| Table | 17-2 | Principal Veins/*113* |
| Table | 18-1 | Blood Types/*121* |
| Table | 18-2 | Blood Tests/*122* |
| Table | 30-1 | Elimination of Waste Products/*197* |
| Table | 35-1 | The Functions of Estrogen and Progesterone/*226* |
| Table | 38-1 | Pituitary Hormones and Their Known Functions/*242* |
| Table | 46-1 | Extrinsic and Intrinsic Eye Muscles/*282* |

# Section 1
# The Body As A Whole

# Unit 1
# INTRODUCTION TO THE STRUCTURAL UNITS

## KEY WORDS

abdominopelvic cavity
anterior
assimilation
bilateral symmetry
caudal
coronal (frontal)
cranial cavity
diaphragm
digestion
distal
dorsal
excretion
gametes
growth
homeostasis
inferior
ingestion
lateral
life function
lower extremities
medial
mediastinum
metabolism
movement
nasal cavity
oral (buccal) cavity
orbital cavity
organism
pericardial cavity
pleura
posterior
proximal
regulation (sensitivity)
reproduction
respiration
sagittal
secretion
spinal cavity
sternum
superior
synthesis
thoracic cavity
transpyloric plane
transtubercular plane
transverse
upper extremities
ventral

## OBJECTIVES

- Identify and discuss the life functions
- Define the chemical processes which maintain life
- Identify the body cavities and the organs they contain
- Define the Key Words that relate to this unit of study

When we examine humans, plants, one-celled organisms, or multicelled organisms, we recognize that all of them have one thing in common: that of being alive.

All living organisms are capable of carrying on life functions. *Life functions* are a series of highly organized and related activities which help living organisms to live, grow, and maintain themselves.

These vital life functions include movement, ingestion, digestion, transport, respiration, synthesis, assimilation, growth, secretion, excretion, regulation (sensitivity) and reproduction, see table 1-1.

## HUMAN DEVELOPMENT

A person is born, grows into maturity, and eventually dies. In the intervening years between birth and death, the body carries on a number of life functions which keep us alive and active. As is true of all living things, each one of us inherits a range of size, form, and a lifespan. We inherit these many characteristics through the gametes from our parents. *Gametes* are the sperm and egg cells.

Living depends upon the constant release of energy in every cell of the body. Powered by the energy that is released from food, the cells are able to maintain their own living condition and thus, the life of human beings.

A complex life form like a human being consists of over fifty thousand billion cells. Early in human development, certain groups of cells become highly specialized for specific functions, like motion or response.

Special cells, grouped according to function, shape, size, and structure are called tissues. Tissues, in turn, form larger functional and structural units known as organs. For

Table 1-1 Review of the Life Functions

| LIFE FUNCTIONS | DEFINITION |
|---|---|
| Movement | The ability of the whole organism — or a part of it — to move |
| Ingestion | The process by which an organism takes in food |
| Digestion | The breakdown of complex food molecules into simpler food molecules |
| Transport | The movement of necessary substances to, into, and around cells, and of cellular products and wastes out and away from cells |
| Respiration | The burning or oxidation of food molecules in a cell to release energy, water and carbon dioxide |
| Synthesis | The combination of simple molecules into more complex molecules to help an organism build new tissue |
| Assimilation | The transformation of digested food molecules into living tissue for growth and self-repair |
| Growth | The enlargement of an organism due to synthesis and assimilation, resulting in an increase in the number and size of its cells |
| Secretion | The formation and release of substances from a cell or structure |
| Excretion | The removal of metabolic waste products from an organism |
| Regulation (sensitivity) | The ability of an organism to respond to its environment so as to maintain a balanced state (homeostasis) |
| Reproduction | The ability of an organism to produce offspring with similar characteristics. This is *essential* for species survival as opposed to individual survival |

example, human skin is an organ made up of epithelial, connective, muscular, and nervous tissue. In much the same way, our kidneys are composed of highly specialized connective and epithelial tissue.

The functional activities of cells that result in growth, repair, energy release, use of food, and secretions are combined under the heading of metabolism. Metabolism consists of two processes which are opposite to each other: anabolism and catabolism. Anabolism means the building up of complex materials from simpler ones. Catabolism is the breaking down and changing of complex substances into simpler ones, with a release of energy. The sum of all the chemical reactions within a cell is, therefore, called *metabolism*.

The proper function and maintenance of the human body depends upon a number of activities. The body must constantly respond to changes in the environment by exchanging substances between its surroundings and its cells. Maintaining the body's cellular environment and function helps to insure regular body functions. Thus optimum cell functioning requires a stable cellular environment (within very narrow limits of acidity, nutrients, oxygen, and temperature). The maintenance of such (optimal) internal environmental conditions is known as *homeostasis*. Human survival ultimately depends on maintenance or restoration of homeostasis.

The organs of the human body do not operate independently. They function interdependently with one another to form a whole, live, functioning organism. Furthermore, some organs are grouped together because they perform a related function. Such a grouping is called an *organ system*. One example is the digestive system, which is composed of all the organs involved in digestion.

# SECTION 1  THE BODY AS A WHOLE

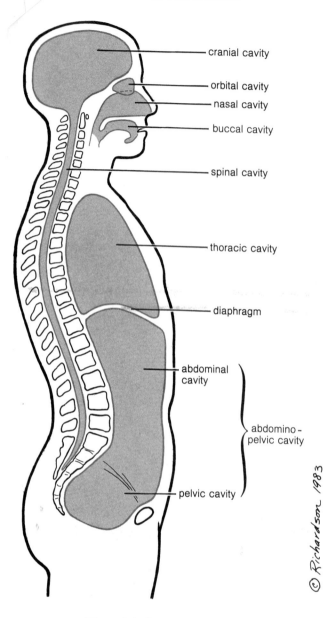

Figure 1-1  Cavities of the body

The circulatory system includes all the organs related to circulation.

## BODY CAVITIES

The organs which comprise most of the nine body systems are organized into several cavities: cranial, spinal, thoracic, and abdominopelvic, figure 1-1. The cranial and spinal cavities are within a larger region known as the dorsal cavity. The thoracic and abdominopelvic cavities are found in the ventral cavity. The dorsal and ventral cavities are the two major body cavities. The *dorsal* (posterior) cavity refers to the back; the *ventral* (anterior) cavity refers to the front or belly side.

The dorsal cavity contains the brain and spinal cord: the brain is in the cranial cavity and the spinal cord is in the spinal cavity, see figure 1-1. The diaphragm divides the ventral cavity into two parts: the upper thoracic and lower abdominopelvic.

The midpoint of the thoracic cavity is known as the *mediastinum*. It is between the lungs and extends from the sternum (breastbone) to the vertebrae of the back. The esophagus, bronchi, lungs, trachea, thymus gland, and heart are located in the mediastinum. The heart itself is contained within a smaller cavity, called the pericardial cavity.

The *thoracic cavity* is further subdivided into two pleural cavities: the left lung is in the left pleural cavity, the right lung is in the right. Each lung is covered with a thin membrane which we call the *pleura*.

The *abdominopelvic cavity* is really one large cavity with no separation between the abdomen and pelvis. In order to avoid confusion, this cavity is usually referred to separately — as the abdominal cavity and the pelvic cavity. The *abdominal cavity* contains the stomach, liver, gallbladder, pancreas, spleen, kidneys, small intestine, appendix, and part of the large intestine, see color plate 1. The kidneys are in the back under the lining of the abdominal cavity.

The urinary bladder, the reproductive organs, the rectum, the remainder of the large intestine, and the appendix are in the *pelvic cavity*.

In order to locate the abdominal and pelvic organs more easily, anatomists have subdivided the abdominopelvic cavity into nine regions and four imaginary planes. The four planes are: two horizontal planes called the transpyloric and the transtubercular and two sagittal planes called the right lateral and the left lateral planes, see figure 1-2.

The nine regions are located in the upper, middle, and lower parts of the abdomen:

- Upper — The right hypochondriac, epigastric, and left hypochondriac regions lie above the upper line (transpyloric plane). Note that the plane crosses the abdomen at the tip of the ninth rib cartilage.

- Middle — The left lumbar, umbilical, and right lumbar regions are located below the transpyloric plane and above an imaginary line (transtubercular plane) that crosses the abdomen at the top of the hip bones.

- Lower — The left iliac, hypogastric, and right iliac regions lie below the transtubercular plane.

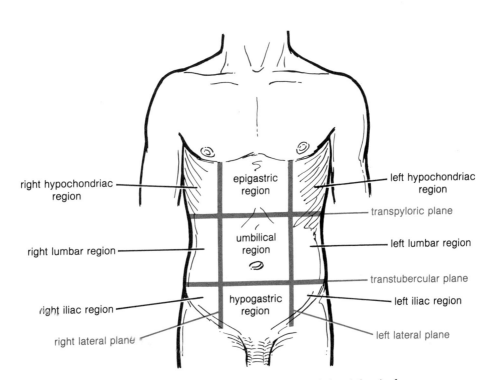

Figure 1-2 Diagram showing the nine regions of the abdominal area

Figure 1-2 will help you understand the various body cavities and anatomical regions of the body.

## Smaller Cavities

In addition to the cranial cavity, the skull also contains several smaller cavities. The eyes, eyeball muscles, optic nerves, and lacrimal (tear) ducts are within the orbital cavity. The nasal cavity contains the parts that form the nose. The oral or buccal cavity encloses the teeth and tongue.

# ANATOMIC TERMINOLOGY

In the study of anatomy and physiology, special words are used to describe the specific location of a structure or organ, or the relative position of one body part to another. Refer to figure 1-3 frequently while studying the following terms:

- *anterior* or *ventral* means "front" or "in front of." For example, the knees are located on the anterior surface of the human body. A ventral hernia may protrude from the front or belly of the abdomen.

- *posterior* or *dorsal* means "back" or "in back of." For example, human shoulder blades are found on the posterior surface of the body. The dorsal aspect of the foot is the back or sole of the foot.

- *cranial* and *caudal* refer to direction; cranial means the head end of the body, caudal means the opposite end. For example, cranial pressure causes headache. Caudal anesthesia is injected in the lower spine.

- *superior* and *inferior* — superior means "upper" or "above another," inferior refers to "lower" or "below another." For example, the heart and lungs are situated superior to the diaphragm, while the intestines are inferior to it.

- *medial* and *lateral* — medial (sometimes called mesial) signifies "toward the midline or median plane of the body," while lateral means "away, or toward the side of the body."

- *proximal* and *distal* — proximal means "toward the point of attachment to the body, or toward the trunk of the body," distal means "away from the point of attachment or origin, or farthest from the trunk." For example, the hand is proximal to the wrist; the elbow is distal to the shoulder. Note: these two words are used primarily to describe the appendages or extremities.

- *sagittal, coronal,* and *transverse planes* — a sagittal or median plane is a lengthwise cut that divides the body into right and left halves. It starts from the middle of the skull, bisecting the breastbone (sternum) and the vertebral column, figure 1-3. The word *sagittal* is derived from the sagittal suture along the skull, which coincides with the midline of the body.

  A coronal (frontal) plane is a vertical cut at right angles to the sagittal plane, dividing the body into anterior and posterior portions. The term *coronal* comes from the coronal suture which runs perpendicular (at a right angle) to the sagittal suture. A *transverse* or cross section is a horizontal cut that divides the body into upper and lower parts.

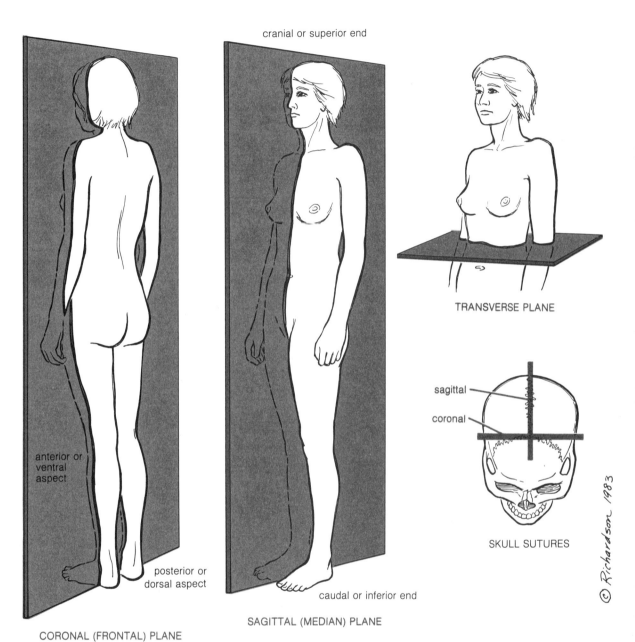

Figure 1-3 Anatomical terms are used to describe body divisions into parts.

# 8 SECTION 1 THE BODY AS A WHOLE

The functioning of complex organisms like human beings includes the combined activities of cells, tissues, organs, and systems. The remainder of this section will cover the structure and functions of cells, specialized groups of cells which are called tissues, and the organs which make up the various systems. In other units of the text, each of the body systems will be discussed in more detail.

## Assignment

Associate each term in column I with its correct description in column II.

| | Column I | Column II |
|---|---|---|
| A | 1. catabolism | a. the sum of all the chemical reactions within the cell |
| | 2. pelvic cavity | |
| | 3. cranial cavity | b. constructive chemical processes which use food to build complex materials of the body |
| | 4. anabolism | |
| | 5. abdominal cavity | |
| | 6. dorsal cavity | c. useful breakdown of food materials resulting in the release of energy |
| | 7. metabolism | |
| | 8. tissue | d. contained within the thoracic cavity |
| | 9. kidneys, ureters and adrenal glands | e. cavity in which the reproductive organs, urinary bladder, and lower part of large intestine are located |
| | 10. heart and lungs | |
| | 11. life function | f. cavity in which the stomach, liver, gallbladder, pancreas, spleen, appendix, cecum, and colon are located |
| | 12. organ system | |
| | | g. the cavity containing both the cranial and spinal cavities |
| | | h. a group of cells which together perform a particular job |
| | | i. portion of the dorsal cavity containing the brain |
| | | j. divides the ventral cavity into two regions |
| | | k. structures located behind the abdominal cavity and under its lining |
| | | l. located in the abdominopelvic cavity |
| | | m. organs grouped together because they have a related function |
| | | n. an activity that a living thing performs to help it live and grow |

# KEY WORDS

absolute zero
active transport
ATP (adenosine triphosphate)
biologist
carrier-molecule complex
cell
cell membrane
cellular respiration
chromatin
chromosome
cristae
cyclosis
cytologist
cytoplasm
deoxyribonucleic acid (DNA)
diffusion
ectoplasm
endoplasm
endoplasmic reticulum (smooth and rough)
enzyme
equilibrium
filtration
Golgi apparatus
hypertonic solution
hypotonic solution
isotonic solution
lipid
lysosome
mitochondria
multicellular
nucleolus
nucleus
organelle
osmolality
osmosis
osmotic pressure
phagocytosis
pinocytic vesicle
pinocytosis
ribosome
selective permeable membrane
species chromosome number
solutes
solvent
unicellular
vacuole

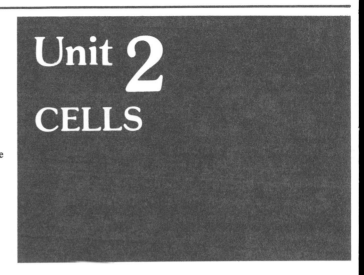

# Unit 2
# CELLS

## OBJECTIVES

- Identify the structure of a typical cell
- Define the function of each component of a typical cell
- Relate the function of cells to the function of the body
- Describe the processes that will transport materials in and out of a cell
- Define the Key Words relating to this unit of study

When a field of grass is seen from a distance it looks just like a solid green carpet. Closer observation, however, shows that it is not a solid mass but is made up of countless separate blades of grass. So it is with the body of a plant or animal; it seems to be a single entity, but when any portion is examined under a microscope it is found to be made up of many small discrete parts. These tiny parts, or units, are called *cells*. All living things, whether plant or animal, unicellular or multicellular, large or small, are composed of cells. A cell is microscopic in size. *The cell is the basic unit of structure and function of all living things.*

Since cells are microscopic, a special unit of measurement is employed to determine their size. This is the micrometer ($\mu$m), or micron ($\mu$). It is used to describe both the size of cells and their cellular components, table 2-1.

To better understand the structure of a cell, let us compare a living entity — such as a

Table 2-1 Units of Length in the Metric System

| |
|---|
| 1 meter = 39.37 inches |
| 1 centimeter (cm) = 1/100 or 0.01 meters |
| 1 millimeter (mm) = 1/1000 or 0.001 meters |
| 1 micrometer ($\mu$m) or micron ($\mu$) = 1/1,000,000 or 0.000001 meters |
| 1 nanometer (nm) = 1/1,000,000,000 or 0.000000001 meters |
| 1 angstrom (Å) = 1/10,000,000,000 or 0.0000000001 meters |

human being — to a house. The many individual cells of this living organism are comparable to the many rooms of a house. Just as each room is bounded by four walls, floor and ceiling, a cell is bounded by a cell membrane. Cells, like rooms, come in a variety of shapes and sizes. Every kind of room or cell has its own unique function. A house can be made up of a single room or many. In much the same fashion, a living thing can be made up of only one cell (*unicellular*), or many cells (*multicellular*).

"Basic" and "typical" are terms used to identify structures common to most living cells.

## THE CELL MEMBRANE

Every *cell* is surrounded by a cell membrane. It is sometimes called a plasma membrane. The cell membrane separates the cell's *cytoplasm* from its external environment and from the neighboring cells. It also regulates the passage or transport of certain molecules into and out of the cell, while preventing the passage of others. This is why the cell membrane is often called a "selective semi-permeable membrane." The cell membrane is made of protein and *lipid* (fatty substance) molecules arranged in a double layer. This arrangement is rather like a sandwich: the lipid molecules are the filling, and the two layers of protein molecules are the slices of bread.

## THE CYTOPLASM

The cytoplasm is a sticky semi-fluid material found between the nucleus and the cell membrane. Cytoplasm may be divided into two layers: an outer layer known as the *ectoplasm* and an inner layer called the *endoplasm*. Chemical analysis of the cytoplasm shows that it consists of proteins, lipids, carbohydrates, minerals, salts, and a great deal of water (70%–90%). Each of these substances varies greatly from one cell to the next and from one organism to the next. The cytoplasm is the background for all the chemical reactions which take place in a cell, such as protein synthesis and cellular respiration. Molecules are transported about the cell by the circular motion of the cytoplasm, (*cyclosis*). Embedded in the cytoplasm are *organelles,* or cell structures that help a cell to function. These are the nucleus, mitochondria, ribosomes, golgi apparatus, endoplasmic reticulum, and the lysosomes.

## THE NUCLEUS

The nucleus is the most important organelle within the cell. It has two vital functions: to control the activities of the cell and to facilitate cell division. This spherical organelle is usually located in or near the center of the cell. Various dyes or stains, like iodine, can be used to make the nucleus stand out. The nucleus stains vividly because it contains *DNA* (deoxyribonucleic acid) and protein. Both readily absorb stains. Surrounding the nucleus is a membrane called the *nuclear membrane.*

The DNA and protein are arranged in a loose and diffuse state called *chromatin.* When the cell is ready to divide, the chromatin condenses to form short, rodlike structures called *chromosomes.* There is a specific number of chromosomes in the nucleus for each species. The number of chromosomes for the human being is 46, or 23 pairs.

When a cell reaches a certain size, it may divide to form two new cells. When this occurs, the nucleus divides first by a process called *mitosis.* During this process, the nuclear material is distributed to each of the two new nuclei. This is followed by division of the

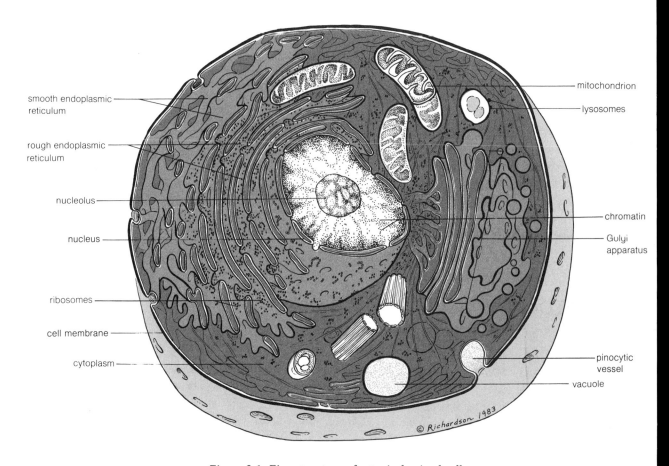

Figure 2-1 Fine structure of a typical animal cell

cytoplasm into two approximately equal parts through the formation of a new membrane between the two nuclei. It is only during the process of nuclear division that the chromosomes can be seen.

Chromosomes are important because they store the hereditary material — deoxyribonucleic acid (DNA) — which is passed on from one generation of cells to the next.

## THE NUCLEOLUS AND THE RIBOSOMES

Within the nucleus are one or more *nucleoli.* Each nucleolus is a small, round body, see figure 2-1. It contains ribosomes made up of ribonucleic acid and protein. The ribosomes can pass from the nucleus through the *nuclear pores* into the cytoplasm. There the ribosomes aid in protein synthesis. They may exist freely in the cytoplasm, be in clusters called *polyribosomes,* or be attached to the walls of the endoplasmic reticulum.

## THE ENDOPLASMIC RETICULUM

Crisscrossing the cellular cytoplasm is a fine network of tubular structures called the *endoplasmic reticulum* (reticulum means "network"). Some of this endoplasmic reticulum connects the nuclear membrane to the cell membrane. Thus it serves as a channel for the transport of materials in and out of the nucleus. Sometimes the endoplasmic reticulum will accumulate large masses of proteins and act as a storage area.

There are two types of endoplasmic reticulum: *rough endoplasmic reticulum* and *smooth endoplasmic reticulum*. Rough endoplasmic reticulum has ribosomes studding the outer membrane. This gives it a coarse appearance. Smooth endoplasmic reticulum has no ribosomes on the outer membrane.

## THE MITOCHONDRIA

All of a cell's energy comes from spherical or rod-shaped organelles called *mitochondria* (singular, mitochondrion; *mito* means thread, *chondrion* means granule). These mitochondria vary in shape and number. There can be as few as a single one in each cell or as many as a thousand or more. Cells that need the most energy have the greatest number of mitochondria. Because they supply the cell's energy, mitochondria are also known as the "powerhouse" of the cell.

The electron microscope identifies the mitochondria as a double-membraned structure: it has an outer membrane and an inner membrane. The inner membrane is folded inward to form shelflike ridges called *cristae*. *Enzymes* are chemicals found in the cristae. These enzymes help the mitochondria to undergo cellular respiration. *Cellular respiration* is a chemical reaction that breaks down carbohydrate, lipid, and protein molecules to release energy, carbon dioxide, and water.

## THE GOLGI APPARATUS

The *Golgi apparatus* was discovered in 1898 by the Italian scientist, Camillo Golgi. It is also called Golgi bodies or the Golgi complex. It is an arrangement of layers of membranes resembling a "stack of pancakes." Scientists believe that this organelle synthesizes carbohydrates and combines them with protein molecules as they pass through the Golgi apparatus. In this way the Golgi apparatus stores and packages secretions for discharge from the cell. It follows logically that these organelles are abundant in the cells of gastric glands, salivary glands and pancreatic glands.

## LYSOSOMES

Lysosomes are oval or spherical bodies found in the cellular cytoplasm. They contain powerful digestive enzymes that digest protein molecules. The lysosome thus helps to digest old, wornout cells, bacteria and foreign matter. If a lysosome should rupture, as sometimes happens, the lysosome will start digesting the cell's proteins, causing it to die. For this reason lysosomes are also known as "suicide bags."

## PINOCYTIC VESICLES

Large molecules like proteins and lipids, which cannot pass through the cell membrane, will enter a cell by way of the pinocytic vesicles. The *pinocytic vesicles* form by having the cell membrane fold inward to form a pocket. Some of the fluid surrounding the cell flows into this pocket. The fluid contains large molecules in solution. The edges of the pocket then close and pinch away from the cell membrane, forming a bubble or *vacuole* in the cytoplasm. The contents of the vacuole are separated from the cytoplasm by a cell membrane. This process by which a cell forms pinocytic vesicles to take in large molecules is called *pinocytosis* or "cell drinking."

## MOVEMENT OF MATERIALS ACROSS CELL MEMBRANES

The cell membrane, aside from housing the cellular organelles, also controls passage of substances into and out of the cell. This is important because a cell must be able to acquire materials from its surrounding medium, after which it either secretes synthesized substances or excretes wastes. The physical processes which control the passage of materials through the cell membrane are: *diffusion, osmosis, filtration, active transport, phagocytosis* and *pinocytosis*. Diffusion, osmosis, and filtration are passive processes, which means they do not need energy in order to function. Active transport, phagocytosis, and pinocytosis are active processes which do require an energy source.

### Diffusion

Diffusion is a physical process whereby molecules of gases, liquids, or solid particles spread or scatter themselves evenly through a medium. When solid particles are dissolved within a fluid, they are known as *solutes*. Diffusion also applies to a slightly different process, where solutes and water pass across a membrane to distribute themselves evenly throughout the two fluids, which remain separated by the membrane. Generally, *molecules move from an area where they are greatly concentrated to an area where they are less concentrated*. The molecules will, eventually, distribute themselves evenly within the space available; when this happens, the molecules are said to be in a state of *equilibrium*, see figure 2-2.

The three common states of matter are gases, liquids, and solids. Molecules will diffuse more quickly in gases and more slowly in solids. Diffusion occurs due to the heat energy of molecules. As a result of this, molecules are always in constant motion, except at *absolute zero* ($-273°C$). In all cases, the movement of molecules increases with an increase in temperature.

A few familiar examples of the rates of diffusion may be helpful. For instance, if one thoroughly saturates a wad of cotton with ammonia and places it in a far corner of a room, the entire room will soon smell of ammonia. Air currents quickly carry the ammonia fumes throughout the room. Another test for diffusion is to place a pair of dye crystals on the bottom of a water-filled beaker. Eventually, they will uniformly permeate and color the water. This diffusion process will take quite a while, especially if no one stirs, shakes, or heats the beaker. In still another test, a dye crystal placed on an ice cube moves even more slowly through the ice. Diffusion of the dye can be accelerated by melting the ice.

The diffusion rate of molecules in the various media (gas, liquid, and solid) depends upon the distances between each molecule and how freely they can move. In a gas, molecules can move more freely and quickly; within a liquid, molecules are more tightly held together. In a solid substance, molecular movement is highly restricted and thus very slow.

Diffusion plays a vital role in permitting molecules to enter and leave a cell. Oxygen diffuses from the bloodstream, where it dwells

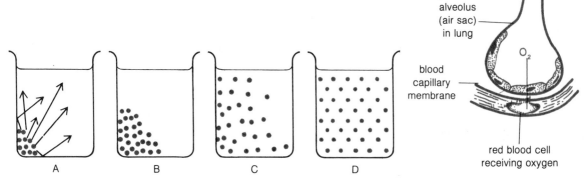

Diffusion:
(A) A small lump of sugar is placed into a beaker of water, its molecules dissolve and begin to diffuse outward. (B&C) The sugar molecules continue to diffuse through the water from an area of greater concentration to an area of lesser concentration. (D) Over a long period of time, the sugar molecules are evenly distributed throughout the water, reaching a state of equilibrium.

Example of diffusion in the human body: Oxygen diffuses from an alveolus in a lung where it is in greater concentration, across the blood capillary membrane, into a red blood cell where it is in lesser concentration.

Figure 2-2 The process of diffusion

in greater concentration. From the bloodstream, the oxygen enters the fluid surrounding a cell, then into the cell itself, where it is far less concentrated. In this manner, the flow of blood through the lungs and bloodstream provides a continuous supply of oxygen to the cells. Once oxygen has entered a cell, it is utilized in metabolic activities.

## Osmosis

Osmosis is the diffusion of water or any other *solvent* molecule through a selective permeable membrane (like the cell membrane). A *selective permeable membrane* is any membrane through which some solutes can diffuse, but others cannot.

Sausage casing is a selective permeable membrane which can be used to substitute for a cell membrane. A solution of salt, sucrose (table sugar), and gelatin is placed into the sausage casing. This mixture is then suspended into a beaker filled with distilled water, figure 2-3. The sausage casing is permeable to water and salt, but not to gelatin and sucrose. Thus only the water and salt molecules can pass through the casing. Eventually more salt molecules will move out because we began with a greater concentration of these molecules inside. At the same time, more water molecules move into the casing, since there were more outside when we began.

This is yet another example of diffusion whereby molecules move from a region of higher concentration to a region of lower concentration. The volume of water increases inside the casing, causing it to expand because of the entry of water molecules. When the number of water molecules entering the casing are equal to the number exiting, an equilibrium has been achieved; the casing will expand no further.

The pressure exerted by the water molecules within the casing at equilibrium is called the *osmotic pressure,* which is expressed as millimeters of mercury (mm Hg).

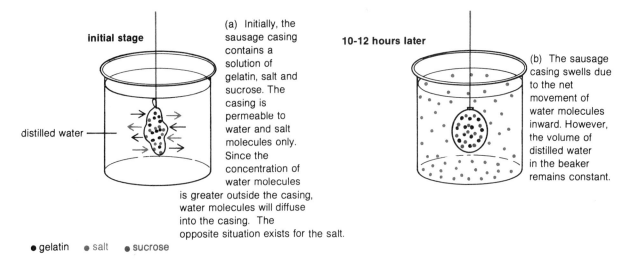

Figure 2-3 Osmosis: the diffusion of water through a selective permeable membrane is illustrated here. (A sausage casing is an example of a selective permeable membrane.)

It is important to remember that every solution has a potential osmotic pressure.

The osmotic pressure of a solution is dependent upon the number of molecules of solute dissolved in a solution. The higher the osmotic pressure (*osmolality*) of a solution, the greater the number of molecules in that solution. And the greater the concentration of molecules, the stronger the "pull" or attraction for water molecules. Simply stated, *water molecules move toward the area of greater osmolality*.

In physiology, the osmotic characteristics of various solutions are determined by the manner in which they affect red blood cells. In other words, the osmolality of a given

**hypertonic solution**

**hypotonic solution**

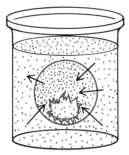

**isotonic solution**

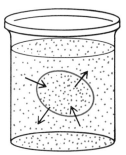

**Hypertonic solution (seawater)**
a red blood cell will shrink and wrinkle up because water molecules are moving out of the cell.
• water molecules

**Hypotonic solution (freshwater)**
a red blood cell will swell and burst because water molecules are moving into the cell.

**Isotonic solution (human blood serum)**
a red blood cell remains unchanged, because the movement of water molecules into and out of the cell are the same.

Figure 2-4 Movement of water molecules in solutions of different osmolalities

solution is compared to that of blood serum. For instance, if a human red blood cell is placed into a solution with the same osmotic pressure as human blood serum, the red blood cell will remain unchanged. This type of solution is known as an *isotonic solution.* In a *hypotonic solution,* the osmolality is lower than that of blood serum, and the red blood cell will swell and burst. This is caused by the water molecules moving into the cell. However, a red blood cell placed inside a *hypertonic solution,* such as seawater (with a higher osmolality than that of blood serum), will shrink and wrinkle up because of the water moving out of the cell, figure 2-4.

## Filtration

Filtration is the movement of solutes and water across a semi-permeable membrane. This results from some mechanical force, such as blood pressure or gravity. The solutes and water move from an area of higher pressure to an area of lower pressure. The size of the membrane pores determines which molecules are to be filtered. Thus filtration allows for the separation of large and small molecules. Such filtration takes place in the kidneys. The process allows larger protein molecules to remain within the body and smaller molecules to be excreted as waste, figure 2-5.

## Active Transport

Active transport is a process whereby molecules move across the cell membrane from an area of lower concentration, against a concentration gradient, to an area of higher concentration. This process requires the high energy chemical compound called ATP (adenosine-triphosphate). The ATP is supplied by the cell membrane.

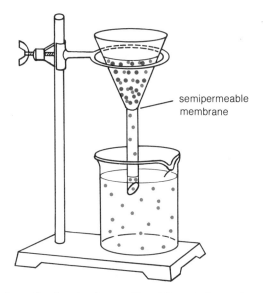

**Filtration:** Small molecules are filtered through the semi-permeable membrane, while the large molecules remain in the funnel.

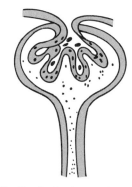

**Example of filtration in the human body:** Glomerulus of kidney, large particles like red blood cells and proteins remain in the blood, and small molecules like urea and water are excreted as a metabolic excretory product—urine.

Figure 2-5 Filtration: a passive transport process

How does active transport work? One theory suggests that a molecule is picked up from the outside of the cell membrane and brought inside by a carrier molecule. Both molecule and carrier are bound together, forming a temporary *carrier-molecule complex.*

This carrier molecule complex shuttles across the cell membrane; the molecule is released at the inner surface of the membrane, from where it enters the cytoplasm. At this point, the carrier acquires energy at the inner surface of the cell membrane. Then it returns to the outer surface of the cell membrane to pick up another molecule for transport. Accordingly, the carrier can also convey molecules in the opposite direction, from the inside to the outside, figure 2-6.

## Phagocytosis

Phagocytosis, or "cell eating," is quite similar to pinocytosis, with an important difference: in pinocytosis, the substances engulfed by the cell membrane are in solution; however, in phagocytosis, the substances engulfed are within particles. Human white blood cells undergo phagocytosis. They will phagocytize bacteria, cell fragments, or even a damaged cell. The particulate substance will be engulfed by an enfolding of the cell membrane to form a vacuole enclosing the material. When the material is completely enclosed within the vacuole, digestive enzymes pour into the vacuole from the cytoplasm to destroy the entrapped substance.

## Pinocytosis

As stated earlier, pinocytosis or "cell drinking" involves the formation of pinocytic vesicles which engulf large molecules in solution. The cell then ingests the nutrient for its own use.

# SPECIALIZATION

There are many kinds of cells of different shapes and sizes. Most of them have the

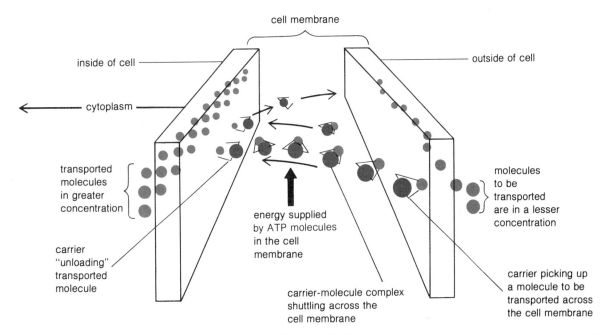

Figure 2-6 Diagram showing the active transport of molecules from an area of lesser concentration to an area of greater concentration, according to one theoretical model

**18** SECTION 1 THE BODY AS A WHOLE

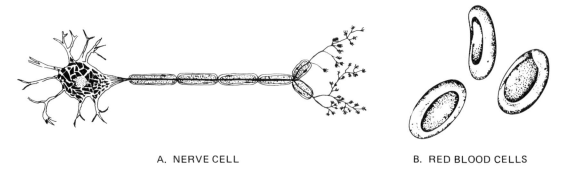

A. NERVE CELL

B. RED BLOOD CELLS

Figure 2-7 Specialized cells

characteristics shown in figure 2-1, which is a generalized diagram of a basic cell. Some of the more specialized types, such as nerve cells and red blood cells, look very different, figure 2-7.

Human beings are composed entirely of cells and the nonliving substances which cells build up around themselves. The interaction of the various parts of the cell within the cellular structure constitutes the life of the cell. These interactions result in the life activities, life processes, or life functions that were discussed in unit 1. However, in complex organisms, groups of cells become specialists in a particular function. Nerve cells, for example, have become specialized in response; red blood cells, in oxygen transport.

Specialized cells may lose the ability to perform some of the other functions, such as reproduction (cell division). Normally, when nerve cells are destroyed or damaged, others cannot be formed to replace them. Specialization also has resulted in an interdependence among cells — certain cells depend on other kinds of cells to aid them in carrying on the total life activities of the organism. In humans, this specialization and interdependence extends to the organs.

## Further Study and Discussion

Complete each of the laboratory exercises under the supervision of your instructor.

## Laboratory Lesson 1

A. **Objective**: To make a slide so as to observe and study the typical cell structure of a cheek cell

B. **Equipment and materials**
- One box of flat-ended toothpicks
- One compound light microscope
- One microscope slide
- One bottle of Lugol's iodine solution
- One cover slip
- Paper towels
- Lens paper
- An eye dropper

C. **Procedure**
1. If necessary, clean the microscope slide and coverslip carefully with the lens paper.
2. Insert the wide, flat end of a toothpick into your mouth. *Gently* scrape the inside of your cheek. (If you have a cold, use your laboratory partner's cheek cells)!
3. Place the wide, flat end of the toothpick against the surface of the microscope slide. Slide the toothpick horizontally across it to make a smear of the cheek cells.
4. Place one or two small drops of Lugol's iodine (solution) on the smear.
5. *Gently* lower the cover slip over a portion of the stained smear.
6. Place your prepared cheek cell slide on the microscope stage. Gently place the stage clips over the slide to secure it.
7. First use the low power objective to focus in on your cells. (Do *not* focus on clumps of cells: only on the separate yellowish cheek cells).
8. Now rotate the low power objective out of its position. Rotate and click the high power objective to focus in on your cells.

D. **Observations**
1. On a separate sheet of paper, draw a cheek cell (as you see it) under low power.
   − Label the structures on the cell.
2. Observing the slide under high power, focus in on the nucleus.
   a. What structure is found within the nucleus?

   b. How many of these are in the nucleus?

3. What is the name of the structure surrounding the entire cheek cell?

4. State what happens to a cheek cell if you move the slide:
   a. to the right?

   b. to the left?

   c. towards you?

   d. away from you?

E. **Conclusions**
   1. Why are some cheek cells clumped together and others separate?

2. Describe the general shape of a cheek cell.

3. Explain why Lugol's iodine solution was added to the cheek cell smear.

4. What is the function of the cell membrane: (a) on the cheek cell, (b) on the nucleus in the cheek cell?

## Laboratory Lesson 2

A. **Objective**: To prepare a slide for observation and study of a human red blood cell

B. **Equipment and Materials**
- One compound microscope
- Four microscope slides
- Two cover slips
- Two sterile disposable lancets
- Sterile absorbent cotton
- 70%–90% alcohol
- One watch glass
- Wright's stain
- Distilled water
- One pipette
- One beaker
- Tap water
- Paper towels

C. **Procedure**

1. Dampen a small wad of sterile absorbent cotton with the alcohol. Swab the tip of the little finger of your left hand.

2. Using a single-wrapped sterile disposable lancet, quickly prick the cleaned area.

3. Immediately squeeze the puncture; a drop of blood should appear.
4. Carefully place the drop of blood half an inch from the edge of a clean glass slide. (Swab the skin puncture with another wad of sterile cotton dampened with alcohol).
5. Use the edge of a second glass slide as a spreader. Hold the spreader slide, with its edge just touching the drop of blood, at a 45 degree angle to the first slide.
6. Allow a few seconds for the drop of blood to spread along the edge of the spreader slide, then push it toward the opposite end of the first slide.

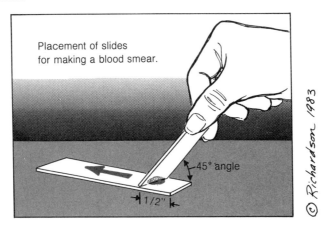

Placement of slides for making a blood smear.

7. Cover the blood smear with a cover slip; examine the slide under the low power lens, then with the high power lens.

**Second part of procedure**: Perform the same exercise, using Wright's stain.

1-6. Repeat steps of preceding exercise in this lesson.
7. Allow two minutes for the blood smear to dry.
8. Rest the slide on top of a watch glass, then add 10 drops of Wright's stain to the slide. Let the excess stain drip into the watch glass.
9. After one minute, add 10 drops of distilled water. Allow 5 minutes for the stain to act.
10. Gently dip the slide in a beaker of tap water to remove any excess stain.
11. Wipe *only* the *lower* surface of the blood smear with paper towels.
12. Cover the blood smear with a cover slip; examine the slide under the low power lens, then with the high power lens.

D. **Observations**

1. How many different types of blood cells did you see?

2. Name the most abundant cell found in blood.

3. Describe the color and shape of these cells.

4. Do these cells contain a nucleus?

5. Did you see a blood cell containing a nucleus?

6. What is the name of this blood cell?

7. Describe the appearance of these cells.

8. Name the smallest blood cell; describe its appearance.

9. How do the red blood cells and the white blood cells compare in number?

E. **Conclusions**
   1. What is the function of the red blood cell?

   2. What is the function of the white blood cell?

   3. What is the function of the platelets?

## Assignment

A. Study the diagram of a typical cell. Enter the names of the structures after the proper numbered callouts, as listed below.

1 _____   8 _____
2 _____   9 _____
3 _____   10 _____
4 _____   11 _____
5 _____   12 _____
6 _____   13 _____
7 _____

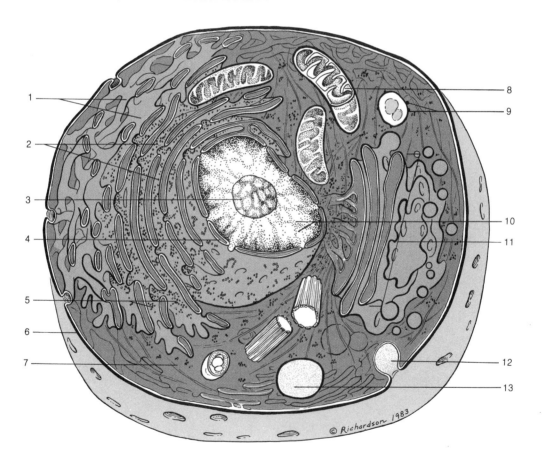

B. Associate each of the terms in column I with its correct description in column II.

| | Column I | Column II |
|---|---|---|
| _____ | 1. nucleus | a. small units of which all plants and animals are made |
| _____ | 2. chromatin material | b. the exposed outer edge of the cell |
| _____ | 3. DNA | c. the process by which cells divide |
| _____ | 4. cytoplasm | d. an example of a specialized cell |
| _____ | 5. nerve | e. the dense inner portion of the cell |
| _____ | 6. reproduction | f. the cell powerhouse from which energy is released |
| _____ | 7. cells | g. the light outer portion of the cell |
| _____ | 8. mitosis | h. an ability lost by some specialized cells |
| _____ | 9. mitochondria | i. the hereditary material within the chromosome |
| _____ | 10. cell membrane | j. cell structure where chromosomes are located |
| | | k. hereditary chemical which transmits traits from one generation to the next. |

# KEY WORDS

| | | |
|---|---|---|
| adipose | elastic | lipid |
| aponeuroses | elastin | neuron |
| areolar | fasciae | osseous |
| calcify | fibrocartilage | pubic bone |
| cardiac | hyaline | pubic symphysis |
| cartilage | intervertebral disc | skeletal |
| collagen | involuntary | sternum |
| conductivity | irritability | tendon |
| costal cartilage | ligament | voluntary |

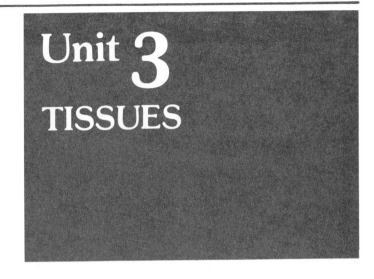

# Unit 3
# TISSUES

## OBJECTIVES

- Describe how cells are organized into tissues
- List the four main types of tissues
- Define the function and location of tissues
- Define the Key Words relating to this unit of study

Multicellular organisms are composed of many different types of cells. Although they are not randomly arranged, each of these cells performs a special function. These millions of cells are grouped according to their similarity in shape, size, structure, and function. Cells so grouped are called *tissues*.

Specialization of cells can be seen in a study of the epithelial cells which make up epithelial tissue. Epithelial cells that cover the body's external and internal surfaces have a typical shape, either columnar, cubical, or platelike. This variation is necessary so the epithelial cells can fit together smoothly in order to line and protect the bodily surface.

Also, muscle cells making up muscle tissue are long and spindle-like so they can contract. Nerve cells (*neurons*) that make up nerve tissue are specialized to carry electrical messages.

Some tissues are comprised of both living cells and various nonliving substances which the cells build up around themselves. This is especially true of the supporting tissues, such as bones and cartilage, figures 3-1 and 3-2.

There are four main types of tissue: (1) epithelial, (2) connective, (3) muscle, and (4) nervous. Each has a specialized structure to perform a particular function. The variations, functions, and locations of each type are described in table 3-1.

**28** SECTION 1 THE BODY AS A WHOLE

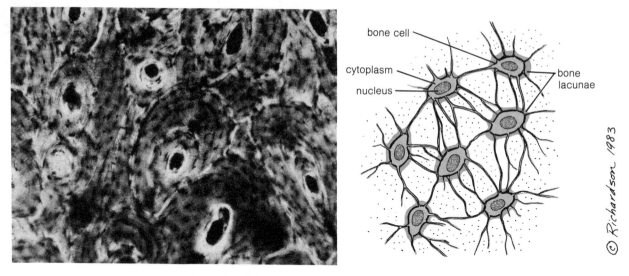

Figure 3-1 Bone tissue (Photograph reprinted, by permission, from Joan G. Creager, *Human Anatomy and Physiology*, 95)

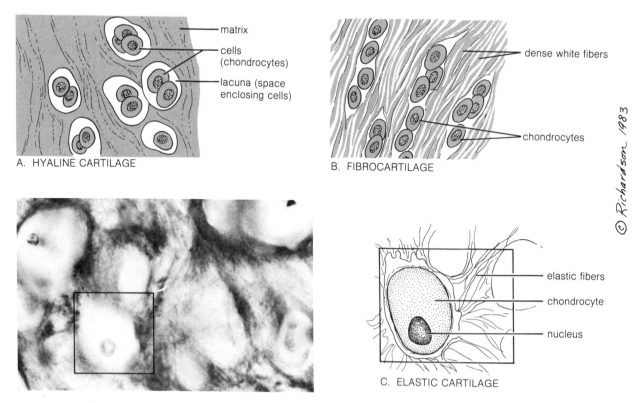

Figure 3-2 Cartilage tissues (Photograph courtesy of Armed Forces Institute of Pathology, negative 71-9216)

Table 3-1 Different Kinds of Human Tissue

| TYPE OF TISSUE | FUNCTION | CHARACTERISTICS AND LOCATION | MORPHOLOGY |
|---|---|---|---|
| I. EPITHELIAL | Cells form a continuous layer covering internal and external body surfaces, provide protection, produce secretions (digestive juices, hormones, perspiration) and regulate the passage of materials across themselves. | | |
| | A. **Covering and lining tissue** <br> These cells can be stratified (layered), ciliated or keratinized. | 1. **Squamous epithelial cells** <br> These are flat, irregularly-shaped cells. They line the heart, blood and lymphatic vessels, body cavities, and alveoli (air sacs) of lungs. The outer layer of the skin is composed of stratified and keratinized squamous epithelial cells. The stratified squamous epithelial cells on the outer skin layer protect the body against microbial invasion. | I-A-1 |
| | | 2. **Cuboidal epithelial cells** <br> These are the cube-shaped cells that line the kidney tubules, and which cover the ovaries and secretory parts of certain glands. | I-A-2 |
| | | 3. **Columnar epithelial cells** <br> Elongated, with the nucleus generally near the bottom and often ciliated on the outer surface. They line the ducts, digestive tract (especially the intestinal and stomach lining), parts of the respiratory tract, and glands. | I-A-3 |

© Richardson 1983

Table 3-1 Different Kinds of Human Tissue (continued)

| TYPE OF TISSUE | FUNCTION | CHARACTERISTICS AND LOCATION | MORPHOLOGY |
|---|---|---|---|
| I. EPITHELIAL (continued) | B. **Glandular or secretory tissue** These cells are specialized to secrete materials like digestive juices, hormones, milk, perspiration and wax. They are columnar or cuboidal shaped. | **Endocrine gland cells** These cells form ductless glands which secrete their substances (hormones) directly into the bloodstream. For instance, the thyroid gland secretes thyroxin, while adrenal glands secrete adrenalin. | I-B: duct (where secretions leave), secretory cells, exocrine (duct) gland cell e.g. sweat and mammary glands |
| II. CONNECTIVE | Cells whose intercellular secretions (matrix) support and connect the many organs and tissues of the body. | Connective tissue is found almost everywhere within the body: bones, cartilage, mucous membranes, muscles, nerves, skin, and all internal organs. | |
| | A. **Adipose tissue** Stores lipid (fat); acts as filler tissue; cushions, supports, and insulates the body. | A type of loose, connective tissue composed of sac-like adipose cells; they are specialized for the storage of fat. Adipose cells are found throughout the body: in the subcutaneous skin layer, around the kidneys, within padding around joints and in the marrow of long bones. | II-A: cytoplasm, collagen fibers, nucleus, vacuole (for fat storage) |
| | B. **Areolar (loose) connective** Surrounds various organs, supports both nerve cells and blood vessels which transport nutrient materials (to cells) and wastes (away from) cells. Areolar tissue also (temporarily) stores glucose, salts and water. | It is composed of a large, semifluid matrix, with many different types of cells and fibers embedded in it. These include fibroblasts (fibrocytes), plasma cells, macrophages, mast cells and various white blood cells. The fibers are bundles of a strong, flexible white fibrous protein called *collagen*, and elastic single fibers of *elastin*. It is found in the epidermis of the skin and in the subcutaneous layer with adipose (fat) cells. | II-B: matrix, reticular fibers, mast cell, collagen fibers, plasma cell, elastic fiber, fibroblast cell, macrophage cell |

© Richardson 1983

Table 3-1 Different Kinds of Human Tissue (continued)

| TYPE OF TISSUE | FUNCTION | CHARACTERISTICS AND LOCATION | MORPHOLOGY |
|---|---|---|---|
| II. CONNECTIVE (continued) | C. **Dense fibrous** This tissue forms ligaments, tendons and aponeuroses. *Ligaments* are strong, flexible bands (or cords) which hold bones firmly together at the joints. *Tendons* are white, glistening bands attaching skeletal muscles to the bones. *Aponeuroses* are flat, wide bands of tissue holding one muscle to another or to the periosteum (bone covering). *Fasciae* are fibrous connective tissue sheets that wrap around muscle bundles to hold them in place. | Dense fibrous tissue is also called white fibrous tissue, since it is made from closely packed white collagen fibers. Fibrous tissue is flexible, but not elastic. It is found in aponeuroses, fasciae, ligaments and tendons. | closely packed collagen fibers; fibroblast cell  II-C |
|  | D. **Supportive** 1. **Bone (osseous) tissue** — Comprises the skeleton of the body, which supports and protects underlying soft tissue parts and organs, and also serves as attachments for skeletal muscles. | Connective tissue whose intercellular matrix is *calcified* by the deposition of mineral salts (like calcium carbonate and calcium phosphate). Calcification of bone imparts great strength. The entire skeleton is composed of bone tissue. | bone cell; cytoplasm; nucleus; bone lacunae  II-D-1 |
|  | 2. **Cartilage** — Provides firm but flexible support for the embryonic skeleton and part of the adult skeleton. a. **Hyaline** — appears as a bluish white, glossy mass. | Hyaline cartilage is found upon articular bone surfaces, and also at the nose tip, bronchi and bronchial tubes. Ribs are joined to the *sternum* (breastbone) by the *costal cartilage*. It is also found in the larynx and the rings in the trachea. | matrix; cells (chondrocytes); lacuna (space enclosing cells)  II-D-2a |
|  | b. **Fibrocartilage** — a strong, flexible, supportive substance; found between bones and wherever great strength (and a degree of rigidity) is needed. | Fibrocartilage is located within *intervertebral discs* and *pubic symphysis* between the *pubic bones*. | dense white fibers; chondrocytes  II-D-2b |

© Richardson 1983

## Table 3-1 Different Kinds of Human Tissue (continued)

| TYPE OF TISSUE | FUNCTION | CHARACTERISTICS AND LOCATION | MORPHOLOGY |
|---|---|---|---|
| II. CONNECTIVE (continued) | D. Supportive (continued)<br>c. Elastic cartilage — the intercellular matrix is embedded with a network of elastic fibers. | Elastic cartilage is located inside the auditory ear tube, external ear, epiglottis, and larynx. | II-D-2c |
| | E. Vascular (liquid blood tissue)<br>1. Blood — Transports nutrient and oxygen molecules to cells, and metabolic wastes away from cells (can be considered as a liquid tissue). | Blood is composed of two major parts: a liquid called plasma, and a solid cellular portion known as blood cells (or corpuscles). The plasma suspends corpuscles, of which there are two major types: *red* blood cells (erythrocytes) and *white* blood cells (leucocytes). A third cellular component (really a cell fragment) is called platelets (thrombocytes). Blood circulates within the blood vessels (arteries, veins and capillaries) and through the heart. | II-E-1 |
| | 2. Lymph — Transports tissue fluid, proteins, fats and other materials from the tissues to the circulatory system. This occurs through a series of tubes called the lymphatic vessels. | Lymph is a fluid made up of water, glucose, protein, fats and salt. The cellular components are lymphocytes and granulocytes. They flow in tubes called lymphatic vessels, which closely parallel the veins and bathe the tissue spaces between cells. | II-E-2 |

Table 3-1 Different Kinds of Human Tissue (continued)

| TYPE OF TISSUE | FUNCTION | CHARACTERISTICS AND LOCATION | MORPHOLOGY |
|---|---|---|---|
| III. MUSCLE | A. **Cardiac** These cells help the heart contract in order to pump blood through and out of the heart. | Cardiac muscle is a striated (having a cross-banding pattern), involuntary (not under conscious control) muscle. It makes up the walls of the heart. | centrally located nucleus; striations; branching of cell; intercalated disc  III-A |
| | B. **Skeletal (striated voluntary)** These muscles are attached to the movable parts of the skeleton. They are capable of rapid, powerful contractions and long states of partially sustained contractions, allowing for voluntary movement. | Skeletal muscle is: *striated* (having transverse bands that run down the length of muscle fiber); *voluntary*, because the muscle is under conscious control; and *skeletal*, since these muscles are attached to the skeleton (bones, tendons and other muscles). | nucleus; myofibrils  III-B |
| | C. **Smooth (nonstriated involuntary)** These provide for involuntary movement. Examples include the movement of materials along the digestive tract, controlling the diameter of blood vessels and the pupil of the eyes. | Smooth muscle is *nonstriated* because it lacks the striations (bands) of skeletal muscles; its movement is *involuntary*. It makes up the walls of the digestive, genitourinary, respiratory tracts, blood vessels and lymphatic vessels. | spindle-shaped cell; nucleus; cells separated from each other  III-C |
| IV. NERVE | **Neuronal** Cells have the ability to react to stimuli. *Irritability* — ability of nerve tissue to respond to environmental changes. *Conductivity* — ability to carry a nerve impulse (message). | Nerve tissue is composed of neurons (nerve cells). Neurons have branches through which various parts of the body are connected and their activities coordinated. They are found in the brain, spinal cord, and nerves. | dendrites; cell body; nucleus; axon; myelin sheath; terminal end branch  IV |

© Richardson 1983

## Further Study and Discussion

- Observe prepared microscope slides of the following tissues: muscle, nerve, epithelial, blood, bone, cartilage, fat, white fibrous, and yellow elastic.

## Assignment

Match each term in column I with its correct description in column II.

| Column I | Column II |
|---|---|
| 1. heart | a. provides protection to the body and produces secretion |
| 2. vascular tissue | b. hardest body tissue providing support |
| 3. epithelial tissue | c. slightly flexible tissue found in the intervertebral disks and the ribs |
| 4. cartilage tissue | d. primarily transports nutrients and wastes |
| 5. differentiation | e. provides for involuntary movements |
| 6. tissue interspaces | f. provides for voluntary movement |
| 7. smooth muscle | g. carries impulses and messages throughout the body |
| 8. bone tissue | h. location of cardiac muscle |
| 9. nervous tissue | i. result of cell specialization |
| 10. skeletal muscle | j. location of most loose fibrous connective tissue |
| | k. fluid in tissue interspaces |
| | l. cell digestion |

## KEY WORDS

anatomy  
division of labor  
organ  
organ system  
physiology  
tissue

# Unit 4
# ORGANS AND SYSTEMS

## OBJECTIVES

- Define an organ
- Define an organ system
- Relate various organs to their respective systems
- Define Key Words relating to this unit of study

An organ is a structure made up of several tissues grouped together to perform a single function. For instance, the stomach is an organ composed of highly specialized vascular, connective, epithelial, muscular, and nerve tissues. All of these tissues function together so as to enable the stomach to undergo digestion and absorption.

The skin which covers our bodies is no mere simple tissue, but a complex organ composed of connective, epithelial, muscular and nervous tissue. These tissues enable the skin to protect the body and remove its wastes (water and inorganic salts), making us sensitive to our environment.

The various organs of the human body do not function separately. Instead, they coordinate their activities to form a complete, functional organism. A group of organs which act together to perform a specific, related function is called an *organ system*.

The *digestive system* has the special function of processing solid food into liquid for absorption into the bloodstream. This organ system includes the mouth, salivary glands, esophagus, stomach, small intestine, liver, pancreas, gallbladder and large intestine. The *circulatory system* transports materials to and from cells. It is comprised of the heart, arteries, veins, capillaries, lymphatic vessels and spleen.

Each of the nine organ systems is highly specialized to perform a specific function; together they coordinate their functions to form a whole, live, functioning organism. This type of specialization is known as *division of labor,* a process which occurs in multicellular organisms.

The systems of the body are the skeletal, muscular, digestive, respiratory, circulatory, reproductive, excretory, endocrine and nervous systems. The functions and organs of each system are shown in table 4-1.

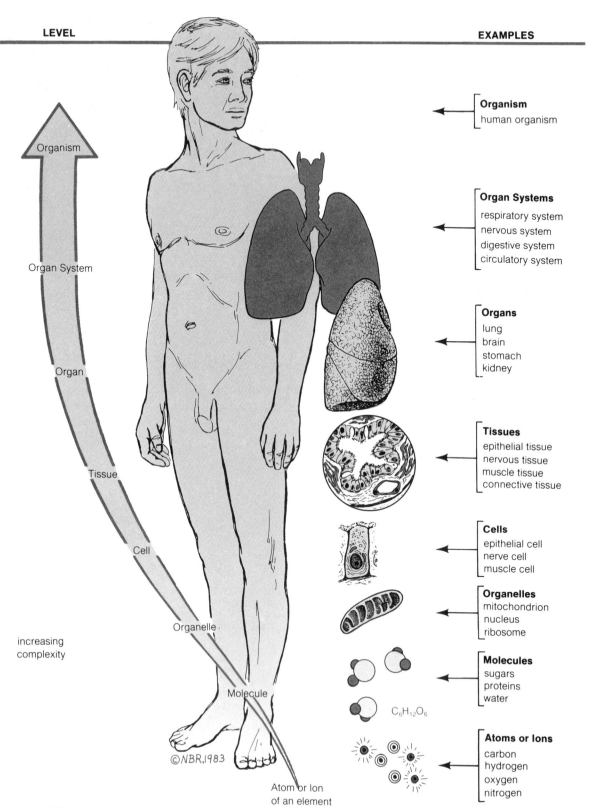

Figure 4-1 Formation of the human organism progresses from different levels of complexity.

Table 4-1 The Nine Body Systems

| SYSTEM | SYSTEM FUNCTIONS | ORGANS |
|---|---|---|
| Skeletal | Gives shape to body; protects delicate parts of body; provides space for attaching muscles; is instrumental in forming blood; stores minerals. | Skull, Spinal Column, Ribs and Sternum, Shoulder Girdle, Upper and Lower Extremities, Pelvic Girdle. |
| Muscular | Determines posture; produces body heat; provides for movement. | Voluntary Muscles (Skeletal) <br> Involuntary Muscles <br> Cardiac Muscle |
| Digestive | Prepares food for absorption and use by body cells through modification of chemical and physical states. | Mouth (salivary glands, teeth, tongue), Pharynx, Esophagus, Stomach, Intestines, Liver, Gallbladder, Pancreas. |
| Respiratory | Acquires oxygen; rids body of carbon dioxide. | Nose, Pharynx, Larynx, Trachea, Bronchi, Lungs. |
| Circulatory | Carries oxygen and nourishment to cells of body; carries waste from cells. | Heart, Arteries, Veins, Capillaries, Lymphatic Vessels, Lymph Nodes, Spleen. |
| Excretory | Removes waste products of metabolism from body. | Skin, Lungs, Kidneys, Bladder, Ureters, Urethra. |
| Nervous | Communicates; controls body activity, coordinates body activity. | Brain, Nerves, Spinal Cord, Ganglia. |
| Endocrine | Manufactures hormones to regulate organ activity. | Glands (ductless): Pituitary, Thyroid, Parathyroid, Pancreas, Adrenal, Gonads (ovaries, testes) |
| Reproductive | Reproduces human beings. | *Male* — Testes, Scrotum, Epididymis, Vas deferens, Seminal vesicles, Ejaculatory duct, Prostate gland, Cowper's gland, Penis, Urethra <br> *Female* — Ovaries, Fallopian tubes, Uterus, Vagina, Bartholin glands, External genitals (vulva), Breasts (mammary glands) |

## Further Study and Discussion

- Identify all the types of tissue present in the arm.
- Discuss how each system is involved in the functioning of the arm.

## Assignment

A. Briefly answer the questions.

1. Explain the relationship of cells to tissues, organs and systems.

# SECTION 1 THE BODY AS A WHOLE

2. Briefly define an organ and an organ system.

3. State the functions of the skeletal system.

4. What is meant by the division of labor between organ systems?

5. List the different types of tissue present in the stomach.

B. Name three organs found in each of the nine body systems.

Skeletal

Excretory

Muscular

Nervous

Digestive

Endocrine

Respiratory

Reproductive

Circulatory

# SELF-EVALUATION

## Section 1 THE BODY AS A WHOLE

A. Complete the following statements.
   1. A group of similar cells which performs one special function is _____.
   2. Three types of muscle tissue are _____, _____, and _____.
   3. Tissue which is found on the surface of the body or lining the body cavities is called _____.
   4. A group of tissues performing one special function is _____.
   5. The part of a cell which directs its activities is _____.
   6. The tissue which provides transportation of materials within the body is _____.
   7. A group of organs which together perform a special function is called _____.
   8. The hardest of the connective tissue adapted to give support and protection is _____.
   9. The tissue which provides for contraction is _____.
   10. Four different kinds of connective tissue are _____, _____, _____, and _____.

B. Classify each of the following according to its main tissue type.
   1. Skin _____
   2. Blood _____
   3. Adipose _____
   4. Heart _____
   5. Brain _____
   6. Skeleton _____
   7. Spinal cord _____
   8. Walls of the stomach _____
   9. Lining of the nose _____
   10. Tendon _____

**40**  SECTION 1  THE BODY AS A WHOLE

C. Name the body system to which each of the following organs belongs.
   1. Brain _____
   2. Adrenal glands _____
   3. Spinal column _____
   4. Voluntary muscles _____
   5. Lungs _____
   6. Heart _____
   7. Kidneys _____
   8. Uterus _____
   9. Stomach _____

D. Label the parts indicated in the following diagram.
   1 _____
   2 _____
   3 _____
   4 _____
   5 _____
   6 _____
   7 _____
   8 _____
   9 _____
   10 _____
   11 _____
   12 _____
   13 _____

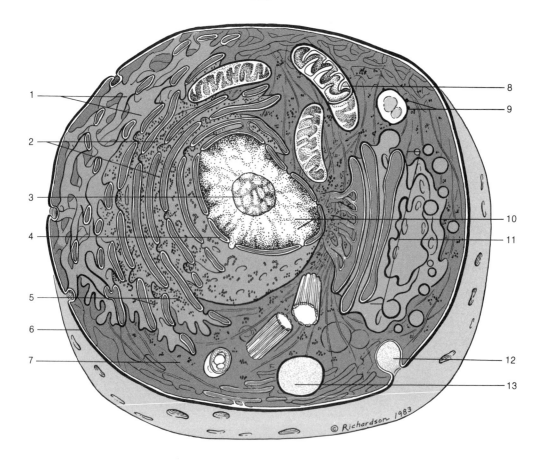

E. Why is the cell called the basic unit of body structure and function?

F. Match each term in column I with its correct definition or description in column II.

| | Column I | Column II |
|---|---|---|
| _____ | 1. abdominal cavity | a. location of brain |
| _____ | 2. abdominopelvic cavity | b. another name for chest |
| _____ | 3. dorsal cavity | c. location of diaphragm |
| _____ | 4. thoracic cavity | d. location of liver |
| _____ | 5. ventral cavity | e. location of urinary bladder |
| _____ | 6. spinal cavity | f. sum total of all life functions |
| _____ | 7. diaphragm | g. location of the spinal cord |
| _____ | 8. anabolism | h. synthesis of complex materials from simpler ones |
| _____ | 9. catabolism | i. tissue separating the thorax from the abdominopelvic cavity |
| _____ | 10. metabolism | j. the breakdown of complex materials into simpler ones |

# Section 2
# The Body Framework

# Unit 5
# INTRODUCTION TO THE SKELETAL SYSTEM

## KEY WORDS

abduction
adduction
atlas
amphiarthroses
articular cartilage
axis
ball and socket joint
bursa
diarthroses
extension
fibrous disc
flexion
gliding joint
herniated (slipped) disc
hinge joint
homeostasis
joint
ligament
pivot joint
pronation
rotation
supination
suture
symphysis pubis
synarthroses
synovial cavity
synovial fluid
synovial membrane
tendon
torso

## OBJECTIVES

- List the main function of bones in the body
- Identify and locate four types of bones
- Name and define the main types of joints
- Name the main types of joint motion
- Define Key Words relating to this unit of study

If you have ever visited a beach, you may have seen a jellyfish floating lightly near the surface. The organs of the jellyfish are buoyed up by the water. But, if a wave should chance to deposit the jellyfish upon the beach, it would collapse into a disorganized mass of tissue. This is because the jellyfish does not possess a supportive framework or *skeleton*. Fortunately, we humans do not suffer such a fate because we have a solid, bony skeleton to support our organs.

The *skeletal system* comprises the bony framework of the body. It is composed of 206 individual bones in the adult; some bones are hinged, while others are fused to one another.

## FUNCTIONS

The skeletal system has four specific functions:

1. **Supports** body structures and provides shape to the body.

2. **Protects** the soft and delicate internal organs. For example, the cranium protects the brain, the inner ear, and parts of the eye. The ribs and breastbone protect the heart and lungs; the vertebral column encases and protects the spinal cord.

3. **Movement** and **anchorage** of muscles. Muscles which are attached to the skeleton are called skeletal muscles. Upon contraction, these muscles exert a pull upon a bone and so move it. In this manner, bones play a vital part in body movement, serving as passively operated levers.

4. **Mineral storage.** Bones are a storage depot for minerals like calcium and phosphorus. In case of inadequate

nutrition, the body is able to draw upon these reserves. For example, when the level of calcium rises above normal in the bloodstream, the excess is stored within the bones. If the level of blood calcium dips below normal, the bone secretes the necessary amount of stored calcium into the bloodstream. In this way the skeletal system helps to maintain blood calcium homeostasis.

## BONE TYPES

Bones are classified as one of four types on the basis of their form, figure 5-1. *Long* bones are found in both upper and lower arms and legs. The bones of the skull are examples of *flat* bones, as are the ribs. *Irregular* bones are represented by bones of the spinal column. The wrist and ankle bones are examples of *short* bones, which appear cube-like in shape.

The bones in the hand are short, making flexible movement possible. The same is true of the irregular bones of the spinal column. The thigh bone is a long bone, needed for support of the strong leg muscles and the weight of the body. The degree of movement at a joint is determined by bone shape and joint structure.

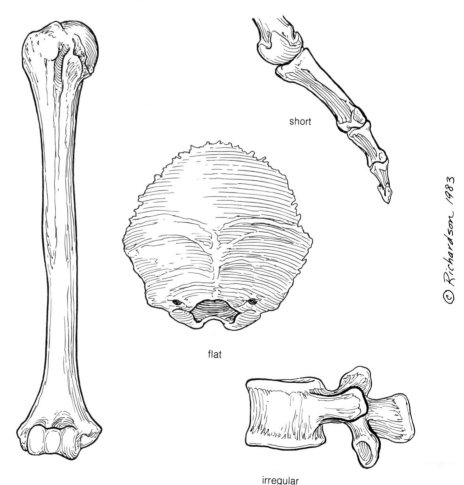

**Figure 5-1** Bone shapes

## JOINTS AND RELATED STRUCTURES

*Joints* are points of contact between two bones. They are classified into three main types according to their degree of movement: *diarthroses* (movable) joints, *amphiarthroses* (partially movable) joints, and *synarthroses* (immovable) joints, figure 5-2.

Most of the joints in our body are diarthroses. They tend to have the same structure. These movable joints consist of three main parts: *articular cartilage,* a *bursa* (joint capsule) and a *synovial* (joint) *cavity*.

When two movable bones meet at a joint, their surfaces do not touch one another. The two *articular* (joint) surfaces are covered with a smooth, slippery cap of cartilage known as *articular cartilage.* This articular cartilage helps to absorb jolts.

Enclosing two articular surfaces of the bone is a tough, fibrous connective tissue capsule called a *bursa.* The bursa securely holds the two articular surfaces while allowing for joint movement. Lining the bursa is a synovial membrane which secretes *synovial fluid* (a lubricating substance) into the synovial cavity

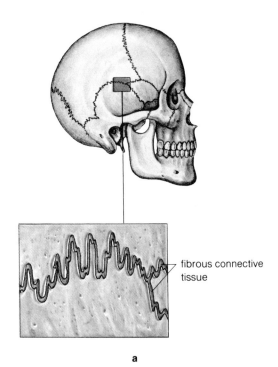
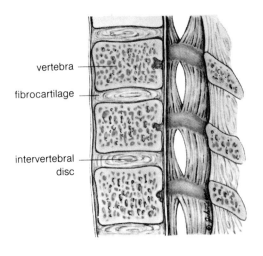
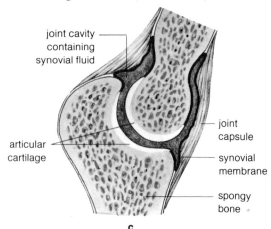

Figure 5-2 Types of joints: (a) a synarthrosis, an immovable fibrous joint (cranial bones); (b) an amphiarthrosis, a slightly movable cartilaginous joint (ribs or vertebra), (c) a diarthrosis, a freely movable hinge or ball-and-socket joint (elbow and hip).

(an area between the two articular cartilages). The synovial fluid reduces the friction of joint movement.

As we advance in age, the joints undergo degenerative changes. The synovial fluid is not secreted as quickly, and the articular cartilaginous surfaces of the two bone ends become ossified. This results in excess bone outgrowths along the joint edges, which tend to stiffen joints, causing inflammation, pain and a decrease in mobility.

There are several types of diarthroses joints: ball-and-socket, hinge, pivot, and gliding joints:

- *Ball-and-Socket joints* allow the greatest freedom of movement. Here, one bone has a ball-shaped head which nestles into a concave socket of the second bone. Our shoulders and hips have ball-and-socket joints.
- *Hinge joints* move in one direction or plane, as in the knees, elbows and outer joints of the fingers.
- *Pivot joints* are those with an extension rotating in a second, arch-shaped bone. The radius and ulna (long bones of the forearm), wrist and ankle are pivot joints. Another example is the joint between the *atlas* (first cervical vertebra in the neck) which supports the head, and the *axis* (second cervical vertebra) which allows the head to rotate.
- *Gliding joints* are those in which nearly flat surfaces glide across each other, as in the vertebrae of the spine. These joints enable the *torso* to bend forward, backward and sideways, as well as rotate.

Between each body of the vertebrae are found *fibrous disks*. At the center of each fibrous disk is a pulpy, elastic material which loses its resiliency with increased usage and/or age. Disks can be compressed by sudden and forceful jolts to the spine. This may cause a disk to protrude from the vertebrae and impinge upon the spinal nerves resulting in extreme pain. Such a condition is known as a *herniated* or "slipped" disk.

*Amphiarthroses* are partially movable joints. These joints have cartilage between their articular surfaces. Two examples are: (1) the attachment of the ribs to the spine and *symphysis pubis,* and (2) the joint between the two pubic bones.

*Synarthroses* are immovable joints connected by tough, fibrous connective tissue. These joints are found in the adult cranium. The bones are fused together in a joint which forms a heavy protective cover for the brain. Such cranial joints are commonly called *sutures.*

*Ligaments* are fibrous bands which connect bones and cartilages and serve as support for muscles. Joints are also bound together by ligaments. *Tendons* are fibrous cords which connect muscles to bones.

## TYPES OF MOTION

Joints can move in many directions, figure 5-3. *Flexion* is the act of bending forward as when the forearm or fingers are bent or flexed. *Extension* means straightening the forearm or fingers. *Abduction* is the movement of an extremity away from the midline (an imaginary line which divides the body from head to foot). *Adduction* is movement toward the midline.

A *rotation* movement allows a bone to move around one central axis. Two rotation movements are pronation and supination. In *pronation,* the forearm turns the hand so the palm is downward or backward. In *supination,* the palm is forward or upward.

## 48 SECTION 2 THE BODY FRAMEWORK

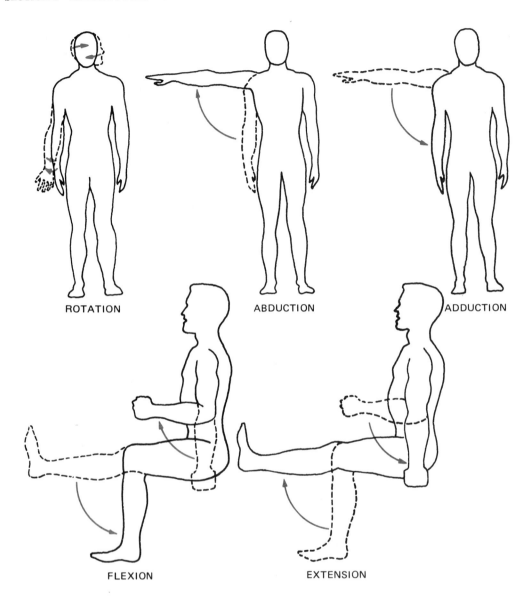

Figure 5-3 Kinds of movements

## Further Study and Discussion

- On a human skeleton model, point out the various joints which are movable, partially movable, and those which are immovable.

- Obtain the legs and wings of a turkey or chicken. Identify the bones, muscles, tendons and ligaments. Note the toughness of the tendons and ligaments. Observe the action of the joints, tendons, and ligaments as the leg is bent and straightened or the wing is spread.
- Discuss the differences in the structure of the male and female skeletons.
- Explain why tendons and ligaments are made of very tough tissue.
- Discuss the functions of the skeletal system.

## Assignment

Select the letter before the word or phrase which most correctly completes the statement.

1. Supination is one type of
   a. extension
   b. abduction
   c. adduction
   d. rotation

2. The bones found in the skull are
   a. irregular bones
   b. flat bones
   c. short bones
   d. long bones

3. The long bones are the site of
   a. storage of fat
   b. hormone secretions
   c. cartilage formation
   d. blood cell formation

4. The cranium protects the
   a. lungs
   b. brain
   c. heart
   d. stomach

5. Pivot joints may be found in the
   a. vertebral column
   b. skull
   c. wrist
   d. shoulder

6. Irregular bones may be found in the
   a. leg
   b. vertebral column
   c. arm
   d. skull

7. Short bones are found in the
   a. leg
   b. arm
   c. vertebral column
   d. hand

8. Immovable joints are found in the
   a. infant's skull
   b. adult cranium
   c. adult spinal column
   d. child's spinal column

9. Flexion means
   a. bending
   b. rotating
   c. extending
   d. abduction

10. The degree of motion at a joint is determined by
    a. the amount of synovial fluid
    b. the number of bursa
    c. the unusual amount of exercise
    d. bone shape and joint structure

# Unit 6
# STRUCTURE AND FORMATION OF BONE

## KEY WORDS

articular cartilage
bone collagen
compact bone
diaphysis
endosteum
epiphyseal cartilage
epiphysis
erythrocyte
fibroblast
fontanel
growth zone
haversian canals
leucocytes
medullary (marrow) canal
ossification
osteoclast
osteocyte
periosteum
spongy bone

## OBJECTIVES

- Explain the formation of bones
- Describe bones with regard to composition and construction
- Relate bone changes to body growth
- Define Key Words relating to this unit of study

Bones are composed of microscopic cells called *osteocytes* (*osteon* — bone; Greek). The osteocytes secrete large amounts of mineral matter. Bone is made up of 35% organic material and 65% inorganic mineral salts.

The organic part derives from a protein called *bone collagen,* a fibrous material. Between these collagenous fibers is a jelly-like material. The organic substances of bone gives it a certain degree of flexibility. The inorganic portion of bone is made from mineral salts like calcium phosphate, calcium carbonate, calcium fluoride, magnesium phosphate, sodium oxide and sodium chloride. These minerals give bone its hardness and durability.

A bony skeleton can be compared to steel-reinforced concrete. The collagenous fibers may be compared to flexible steel supports, and mineral salts to concrete. When tension is applied to a bone, the flexible, organic material prevents bone damage, while the mineral elements resist crushing under pressure.

## BONE FORMATION

The embryonic skeleton is initially composed of collagenous protein fibers secreted by the *fibroblasts* (primitive embryonic cells). Later on, during embryonic development, cartilage is deposited between the fibers. At this stage, the embryo's skeleton consists of collagenous protein fibers and hyaline (clear) cartilage. During the eighth week of embryonic development, *ossification* begins. That is, mineral matter starts to replace previously formed cartilage, creating bone. Infant bones are very soft and pliable because of incomplete ossification at birth. A familiar example is the soft spot on a baby's head, the *fontanel,* see color plate 2. The

51

## 52 SECTION 2 THE BODY FRAMEWORK

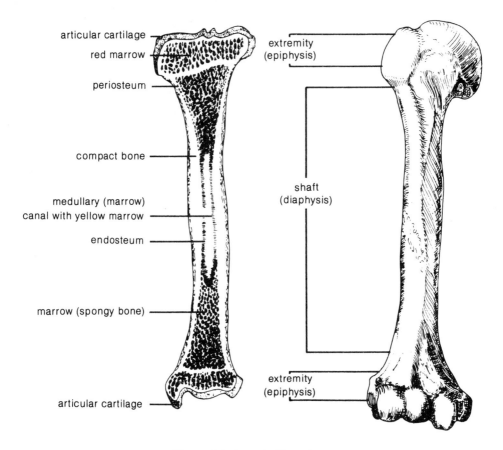

Figure 6-1 A typical long bone

bone has not yet been formed there, although it will become hardened later. Ossification due to mineral deposits continues through childhood. As bones ossify, they become hard and more capable of bearing weight.

## STRUCTURE

A typical long bone is composed of a shaft, or *diaphysis*. This is a hollow cylinder of hard, compact bone. It is what makes a long bone strong and hard yet light enough for movement. At the ends (extremes) of the diaphysis are the *epiphysis*, figure 6-1.

In the center of the shaft is the broad *medullary canal*. This is filled with yellow bone marrow, mostly made of fat cells. The marrow also contains many blood vessels and some cells which form white blood cells, called *leucocytes*. The yellow marrow functions as a fat storage center. The marrow canal is lined and the cavity kept intact by the *endosteum*.

The medullary canal is surrounded by *compact* or hard bone. *Haversian canals* branch into the compact bone. They carry blood vessels which nourish the *osteocytes*, or bone cells. Where less strength is needed in the bone, some of the hard bone is dissolved away leaving *spongy* bone.

The ends of the long bones contain the red marrow where some red blood cells called *erythrocytes* and some white blood cells are made. The outside of the bone is covered with the *periosteum,* a tough fibrous tissue which contains blood vessels, lymph vessels and nerves. The periosteum is necessary for bone growth, repair and nutrition.

Covering the epiphysis is a thin layer of cartilage known as the *articular cartilage.* This cartilage acts as a shock absorber between two bones that meet to form a joint.

## GROWTH

Bones grow in length and ossify from the center of the diaphysis toward the epiphyseal extremities. Using a long bone by way of example, it will grow lengthwise in an area called the *growth zone.* Ossification occurs here, causing the bone to lengthen; this causes the epiphyses to grow away from the middle of the diaphysis. It is a sensible growth process, since it doesn't interfere with the articulation between two bones.

A bone increases its circumference by the addition of more bone to the outer surface of the diaphysis. As girth increases, bone material is being dissolved from the central part of the diaphysis. This forms an internal cavity called the *marrow cavity,* or *medullary canal.* The medullary canal gets larger as the diameter of the bone increases.

The dissolution of bone from the medullary canal results from the action of cells called osteoclasts. *Osteoclasts* are very large bone cells which secrete enzymes. These enzymes digest the bony material, splitting the bone minerals and enabling them to be absorbed by the surrounding fluid. The medullary canal eventually fills with yellow marrow and cells that will produce white blood cells.

The length of a bone shaft continues to grow until all the epiphyseal cartilage is ossified. At this point, bone growth stops. This fact is helpful in determining further growth in a child. First, an X ray of the child's wrist is taken. If some epiphyseal cartilage remains, there will be further growth. If there is no epiphyseal cartilage left, the child has reached his or her full stature (height).

The average growth in females continues to about 18 years, in males to approximately 20 or 21 years. However, new bone growth can occur in a broken bone at any time. Bone cells near the site of a fracture become active, secreting large amounts of new bone within a relatively short time. Bone healing proceeds quickly and efficiently in youth. The process can be helped along when the fractured ends are properly aligned and immobilized by a cast, splint or by the insertion of a bone pin.

---

## Further Study and Discussion

- Obtain a beef bone from the butcher shop. Have it sawed through the center — lengthwise so that the inner portion of the bone is visible. Identify each part of the bone.

- Discuss how the periosteum is involved in the growth of a bone.

## Assignment

A. Briefly answer the following questions.

1. Describe the composition of bone.

2. Explain the difference between compact bone and spongy bone.

3. Explain the function of the red marrow of the bone.

4. Explain the function of the yellow marrow of the bone and what it is made from.

5. Describe the function of the Haversian canals.

6. Why do infant bones tend to be soft and pliable?

UNIT 6 STRUCTURE AND FORMATION OF BONE 55

7. How may it be determined whether or not a child will have further bone growth?

8. Discuss the pattern of bone growth.

9. How can the process of bone healing be helped along?

B. Match each term in column I with its correct description in column II.

| Column I | Column II |
|---|---|
| \_\_\_\_ 1. mineral matter | a. dietary elements which furnish cells with necessary materials to manufacture mineral matter |
| \_\_\_\_ 2. ossification | |
| \_\_\_\_ 3. fontanel | |
| \_\_\_\_ 4. endosteum | b. center of the bone shaft |
| \_\_\_\_ 5. calcium and phosphorus | c. part of bone containing yellow marrow, blood vessels, and some cells which form white blood cells |
| \_\_\_\_ 6. epiphysis | |
| \_\_\_\_ 7. periosteum | d. stage of development when bones begin to form |
| \_\_\_\_ 8. bone marrow | |
| \_\_\_\_ 9. medullary canal | e. the process of mineral deposition and bone cell growth |
| \_\_\_\_ 10. early embryonic period | f. another term for sodium chloride |
| | g. area in infant skull where bone has not yet formed |
| | h. lining of the bone marrow canal |
| | i. elements which make bones hard and durable |
| | j. end structure of a long bone |
| | k. bone cells or osteocytes |
| | l. bone covering which contains blood vessels, lymph vessels, and nerves |

# Unit 7

# PARTS OF THE SKELETON

## KEY WORDS

| | | |
|---|---|---|
| acetabulum | hyoid bone | quadripedal |
| appendicular skeleton | ilium | radius |
| | innominate bones | sacroiliac joint |
| articular process | intervertebral disc | scapula |
| axial skeleton | ischium | sesamoid bone |
| bipedal | manubrium | spinous process |
| calcaneous | metacarpal | sphenoid |
| clavicle | metatarsal | suture |
| costal cartilage | occipital bone | talus |
| ethmoid | olecranon process | tarsus |
| femur | ossa carpi | temporal bone |
| fibula | paranasal sinus | tibia |
| foramen | parietal bone | transverse process |
| frontal bone | patella | ulna |
| glenoid fossa | phalange | vertebrae |
| humerus | pubis | xiphoid process |

## OBJECTIVES

- Name the components of the two main parts of the human skeleton
- Describe the functions of the main bone structures
- Locate the bones in the human skeleton
- Define Key Words that relate to this unit of study

---

The skeletal system is comprised of two main parts. The *axial skeleton* consists of the skull, spinal column, ribs, breastbone and hyoid bone. The hyoid bone is a U-shaped bone in the neck. The tongue is attached to it. The *appendicular skeleton* includes the upper extremities: shoulder girdles, arms, wrists, hands, and the lower extremities: hip girdle, legs, ankles and feet, figure 7-1.

## AXIAL SKELETON

The skull is composed of the cranium and facial bones. The cranium houses and protects the delicate brain, while the facial bones guard and support the eyes, ears, nose and mouth. Some of the facial bones, such as the nasal bones, are made of bone and cartilage. For example, the upper part of the nose (bridge) is bone, while the lower part is cartilage.

Cranial bones are thin and slightly curved. During infancy, these bones are held snugly together by an irregular band of connective tissue called a *suture*. As the child grows, this connective tissue ossifies and turns into hard bone. Thus the cranium becomes a highly efficient, dome-shaped shield for the brain. The dome shape affords better protection than a flat surface, deflecting blows directed toward the head. However, it is not invulnerable and a particularly hard blow may fracture it. This can lead to a concussion: if the bone is depressed, serious injury to brain tissue may result. A depressed fracture may require surgery to relieve the pressure from the brain.

Collectively, there are twenty-two bones in the skull. Eight bones are in the cranium: frontal bone, left and right parietal, occipital, left and right temporal, sphenoid, and ethmoid. The remaining fourteen are facial bones. Of

UNIT 7 PARTS OF THE SKELETON 57

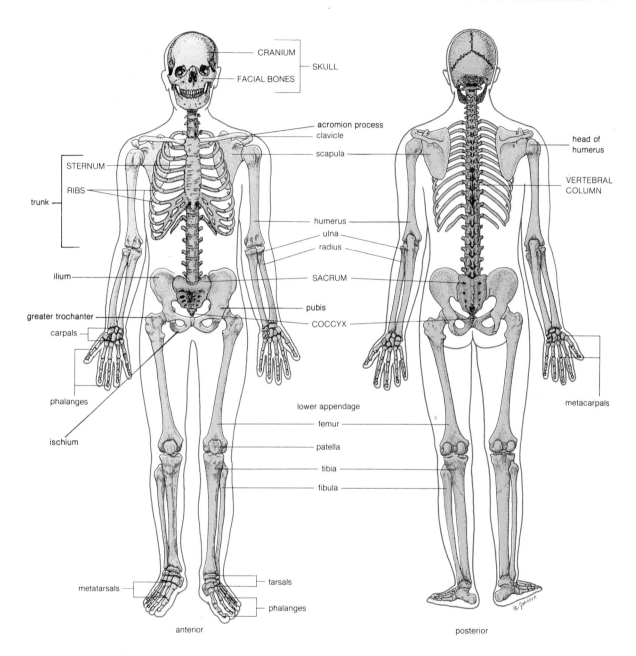

Figure 7-1 Bones of the skeleton

these, the mandible (lower jaw) is the largest and aids in chewing. The smallest bones are the three within the ear: the hammer, anvil, and stirrup; they play a vital role in hearing, see color plates 3 and 4.

The skull contains large spaces (cavities) within the facial bones, referred to as *paranasal sinuses*. These sinuses are lined with mucous membranes. When a person suffers from a cold, flu or hayfever, the membranes become inflamed and swollen, producing a copious amount of mucus. This may lead to sinus pain and a "stuffy" nasal sensation.

## Spine

The spine, or vertebral column, is strong and flexible. It supports the head and provides for the attachment of the ribs. The spine also encloses the spinal cord of the nervous system.

The spine consists of small bones called *vertebrae* which are separated from each other by pads of cartilage tissue called *invertebral disks*. The vertebral column is divided into five sections from the first to the last vertebrae: *cervical* (neck), *thoracic* (chest), *lumbar* (lower back), *sacral* (hip) and *coccyx* (tail), figure 7-2. The first vertebra is the atlas, which supports the skull. The second is called the axis: it makes rotation of the skull possible as the atlas turns upon the axis.

There are 33 separate vertebrae within the developing embryo. Before birth, several will fuse together, leaving only 26. These include seven cervical, twelve thoracic, five fused lumbar vertebrae, five fused sacral vertebrae, and four coccygeal vertebrae, figure 7-3.

When you study a model of the human skeleton, you will notice that the spine is curved instead of straight. A curved spine has more strength than a straight one would

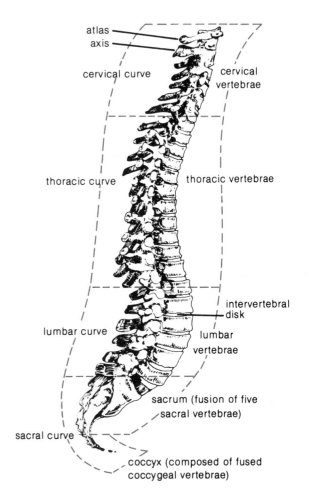

Figure 7-2 Lateral view of the spine

have. Furthermore, its shape provides the proper balance for human *bipedal* (two-footed) posture, as opposed to the *quadripedal* (four-footed) posture of most animals. Before birth, the thoracic and sacral regions are convex curves. As the infant learns to hold up its head, the cervical region becomes concave. When the child learns to stand, the lumbar region also becomes concave. This completes the four curves of a normal, adult human spine.

UNIT 7 PARTS OF THE SKELETON 59

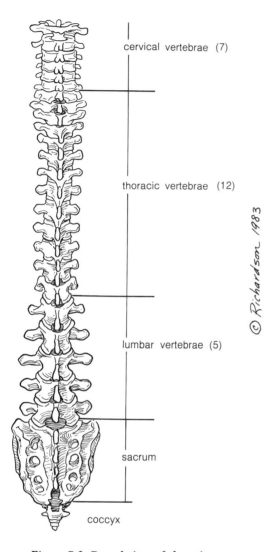

Figure 7-3 Dorsal view of the spine

A typical vertebra, as seen in figure 7-4, contains three basic parts: body, foramen and (several) processes. The large, solid part of the vertebra is known as the *body;* the central opening for the spinal cord is called the *foramen.* Above the foramen protrude two wing-like bony structures called *transverse processes.* The roof of the foramen contains the *spinous process* (spine) and the *articular processes.*

Ribs and Sternum

The thoracic area of the body is protected and supported by the thoracic vertebrae, ribs, and sternum.

The breastbone (sternum) is divided into three parts: the upper region (manubrium), the body, and a lower cartilaginous part called the xiphoid process. Attached to each side of the upper region of the sternum, by means of ligaments, are the two *clavicles* (collar bones).

Seven pairs of *costal cartilages* join seven pairs of ribs directly to the sternum. These are known as *true ribs.* The human body contains twenty-four pairs of ribs. The first seven pairs are true ribs. The next three pairs are "false ribs" because their costal cartilages are attached to the seventh rib instead of directly to the sternum. Finally, the last two pairs of ribs, connected neither to the costal cartilages nor the sternum, are floating ribs, figure 7-5.

## THE APPENDICULAR SKELETON

The appendicular skeleton includes the bones in the upper and lower extremities; the axial includes bones of the head and trunk. There are 126 bones in the appendicular skeleton.

Shoulder Girdle

The shoulder girdle consists of four bones: two curved clavicles and two triangular scapulae (shoulder bones). Using a model of the human skeleton, we observe two broad, flat triangular surfaces (scapulae) on the upper posterior surface. They permit the attachment of muscles which assist in arm movement, while also serving as a place of attachment for the arms. The two clavicles, attached at one end to the scapulae and at the

**60** SECTION 2 THE BODY FRAMEWORK

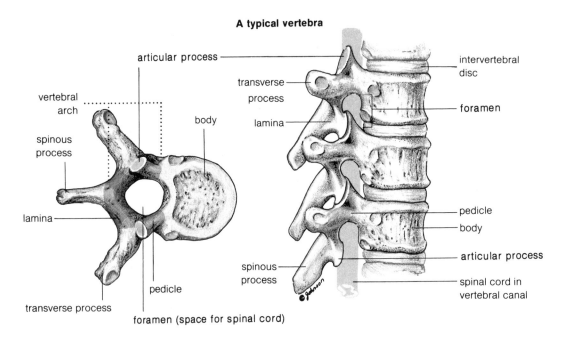

Figure 7-4 Vertebrae comparison

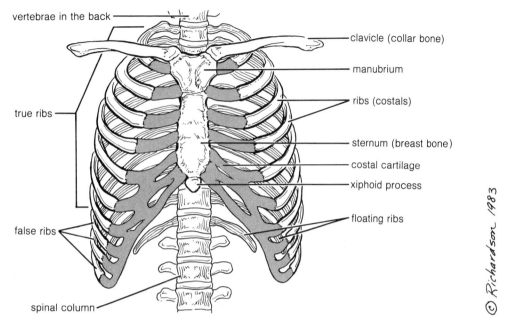

**Figure 7-5** Ribs and breastbone

other to the sternum, help to brace the shoulders and prevent excessive forward motion.

## Arm

The bone structure of the arm consists of the humerus, the radius, and the ulna. The *humerus* is located in the upper arm and the *radius* and *ulna* in the forearm.

The humerus, the only bone in the upper arm, is the second largest bone in the body. The upper end of the humerus has a smooth, round surface called the head, which articulates with the scapula. The upper humerus is attached to the scapula socket (*glenoid fossa*) by muscles and ligaments. These muscles are the biceps and triceps brachii.

The forearm is composed of two bones: the radius and the ulna. The *radius* is the bone running up the thumb side of the forearm. Its name derives from the fact that it can rotate around the ulna. This is an important characteristic, permitting the hand to rotate freely and with great flexibility. The ulna, by contrast, is far more limited. It is the largest bone in the forearm; at its upper end, it produces a projection called the *olecranon process,* forming the elbow.

## Hand

The human hand is a remarkable piece of skeletal engineering and dexterity. It contains more bones for its size than any other part of the body. Collectively, the hand has twenty-seven bones and an opposable thumb, figure 7-6.

The wrist bone, or *ossa carpi,* is comprised of eight small bones arranged in two rows. They are held together by ligaments which permit sufficient movement to allow the wrist a great deal of mobility and flexion. However, there is very little lateral (side) movement of these carpal bones. On the palmar side of the hand are attached a number of short muscles

## 62 SECTION 2 THE BODY FRAMEWORK

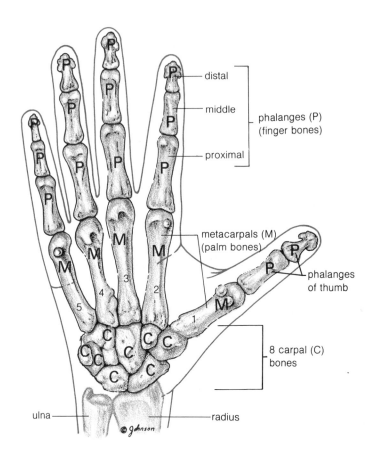

Figure 7-6 Diagram showing the 27 bones of the right hand — dorsal view

which supply mobility to the little finger and thumb.

The hand is composed of two parts: the palmar surface with five *metacarpal bones*, and five fingers comprised of fourteen *phalanges* (singular, phalanx). Each finger, except for the thumb, has three phalanges, whereas the thumb has two. There are hinge joints between each phalanx, allowing the fingers to be bent easily. The thumb is the most flexible finger because the end of the metacarpal bone is more rounded, and there are muscles attached to it from the hand itself. Thus the thumb can be extended across the palm of the hand. Only man and other primates possess such a digit known as an *opposable thumb*.

## Pelvic Girdle

In youth, the pelvic girdle (*innominate bones*) consists of three bones. Found on either side of the midline of the body, the innominate bones include the *ilium* (plural, *ilia*), the *ischium* (plural, *ischia*) and the *pubis* (plural, *pubes*). However, these bones eventually fuse with the sacrum to form a bowl-shaped structure called the pelvic girdle. Eventually these two sets of innominate bones

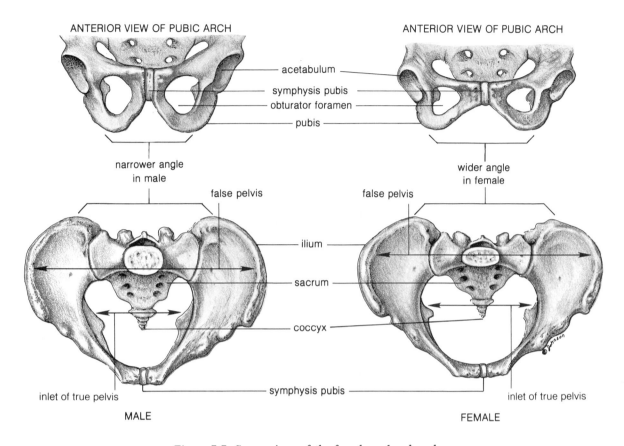

Figure 7-7 Comparison of the female and male pelves

form a joint with the bones in front, called the *symphysis pubis* and with the sacrum in back, as the *sacroiliac joint*.

The pelvic girdle serves as an area of attachment for the bones and muscles of the leg. It also provides support for the viscera (soft organs) of the lower abdominal region. There is an obvious anatomical difference between the male and female pelvis. The female pelvis is much wider than that of the male. This is necessary for childbearing (pregnancy) and childbirth. In addition, the *pelvic inlet* is wider in the female, and the pelvic bones are lighter and smoother than those of the male, figure 7-7.

Upper Leg

The upper leg contains the longest and strongest bone in the body, the thigh bone or *femur*. The upper part of the femur has a smooth rounded head. It fits neatly into a cavity of the ilium known as the *acetabulum*, forming a ball-and-socket joint. The femur is an amazingly strong bone. A direct compressible force applied to the *top* of the femur of from 15,000–19,000 pounds per square inch is required to break it. How then, do fractures occur? In the event of a side blow or twisting motion, a few hundred pounds of pressure per square inch is enough to break the femur.

## Lower Leg

The lower leg consists of two bones: the *tibia* and the *fibula*. The tibia is the largest of the two lower leg bones. The *patella* (kneecap) is found in front of the knee joint. It is a flat, triangular, sesamoid bone. The patella is formed in the tendons of the large muscle in front of the femur (quadriceps femoris). In females, it appears at around two or three years of age; in males, at about six. The patella, attached to the tibia by a ligament, ossifies as early as puberty. Surrounding the patella are four bursae, which serve to cushion the knee joint.

## The Ankle

The ankle (*tarsus*) contains seven tarsal bones. These bones provide a connection between the foot and leg bones. The largest ankle bone is the heel bone or *calcaneous*. The tibia and fibula articulate with a broad tarsal bone called the *talus*. Ankle movement is a sliding motion, allowing the foot to extend and flex when walking.

## The Foot

The foot has five metatarsal bones which are somewhat comparable to the metacarpals

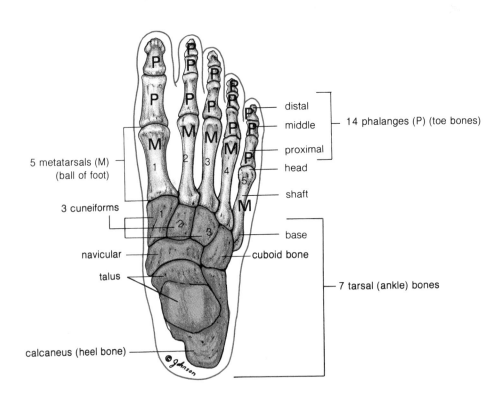

Figure 7-8 Dorsal view of the right foot and its 26 bones

of the hand. But, there is an important difference between the metatarsals and the metacarpals within the palm of the hand. The metatarsal and tarsal bones are arranged to form two distinct arches, which of course are not found in the palm. One arch runs longitudinally from the calcaneus to the heads of the metatarsals; it is called the *longitudinal arch*. The other, which lies perpendicular to the longitudinal arch in the metatarsal region, is known as the *transverse arch*. Strong ligaments and leg muscle tendons help to hold the foot bones in place to form those two arches. In turn, arches strengthen the foot and provide flexibility and springiness to the stride. In certain cases, these arches may "fall" due to weak foot ligaments and tendons. Then downward pressure by the weight of the body slowly flattens them, causing "fallen arches" or "flatfeet." Flatfeet cause a good deal of stress and strain on the foot muscles, leading to pain and fatigue. Factors which may lead to flatfeet include improper prenatal nutrition, dietary or hormonal imbalances, fatigue, overweight, poor posture, and shoes which do not fit properly.

The toes are similar in composition to the fingers. There are three phalanges in each, with the exception of the big toe which has only two. Since the big toe is not opposable like the thumb, it cannot be brought across the sole. There are a total of fourteen phalanges in each foot, figure 7-8.

## Further Study and Discussion

- Using a model of the human skeleton, identify the following structures:
    - the cranium, the zygomatic arch, maxilla, and the frontal bone
    - shoulder girdle, clavicle and scapula
    - sternum, manubrium and xiphoid process
    - pelvic girdle, ilium, ischium and the pubis
    - spinal column, the atlas, the axis, the cervical vertebra, the thoracic vertebra, the lumbar vertebra, the sacrum and the coccyx
    - arms, humerus, radius and the ulna
    - legs, femur, patella, fibula and the tibia
    - foot, calcaneous, talus, metatarsals and the phalanges
- Find out why the two lowest pairs of ribs are called floating ribs
- Explain why the first vertebra in the spinal column is called the atlas and why it has no body like other vertebra do.

## Assignment

A. Complete the following statements.
   1. The two main parts of the skeletal system are the _____ skeleton and the _____ skeleton.

2. Two main areas of the skull consist of the _____ and the _____.
3. The largest of the fourteen facial bones is the _____.
4. The three tiny bones of the ear which assist in the hearing function are the _____, _____, and _____.
5. The spinal or vertebral column consists of small bones separated from each other by pads of _____ called _____ _____.
6. The odontoid process is an important structure of the second vertebra, the _____.
7. the flat bone lying between the ribs in the front of the chest is the _____.
8. The bones which form the pelvic girdle are called the _____ bones.
9. The individual bones which form the pelvis are the _____, _____, _____, _____ and _____.
10. The largest bone in the body is the _____ or _____ bone.
11. The bones of the skull are the _____, _____, _____, and _____ bones.
12. The sutures of the skull are the _____, _____, and _____ sutures.
13. The spinal column has five main vertebral sections, the _____, _____, _____, _____, and _____.
14. The first vertebra which supports the skull is the _____.
15. The odontoid process forms a pivot upon which the _____ vertebra rotates.

B. Answer the following questions.
   1. What is the function of the hyoid bone?

2. Explain why the cranium is dome-shaped.

3. Why is the spinal column curved instead of straight?

4. What is the function of a foramen?

5. Describe the anatomical differences between the male and female pelvic girdle.

6. What does the term "opposable thumb" mean?

7. Explain what is meant by "flatfeet." How does this condition occur?

# KEY WORDS

ankylosis
arthritis
bunion
bursitis
clubfoot (talipes)
comminuted
   fracture
compound fracture
cranium
dislocation
fracture
gout (gouty
   arthritis)
greenstick fracture
intervertebral disk
kyphosis
lordosis
lumbar vertebrae
microcephalus
osteoarthritis
osteoporosis
reduction
rickets
rheumatoid
   arthritis
sacral vertebrae
scoliosis
simple fracture
spina bifida
sprain
subluxation
thoracic
   vertebrae

# Unit 8
# REPRESENTATIVE DISORDERS OF THE BONES AND JOINTS

## OBJECTIVES

- Define four types of bone fractures
- Identify common bone and joint injuries
- Identify common bone and joint disorders
- Define the Key Words that relate to this unit of study

---

The most common injury to a bone is a *fracture,* or break. When this occurs, there is swelling due to injury and bleeding tissues. The process of restoring the fractured bone to its original position is known as *reduction.* A cast is applied to hold the fracture in place and at rest. Healing takes place and the bone knits, or grows together again. The following outline identifies the common types of fractures, figure 8-1.

- Simple — The bone is broken, but the broken ends *do not* pierce through the skin forming an external wound.
- Compound — This is the most serious type of fracture, where the broken bone ends pierce and protrude through the skin. This can cause infection of the bone and of the neighboring tissues.
- Greenstick — Here we have the simplest type of fracture. The bone is partly bent, but it never completely separates. The break is similar to that of a young, sap-filled woodstick, where the fibers separate lengthwise when bent. Such fractures are common among children because their bones contain flexible cartilage.
- Comminuted — The bone is splintered or broken into many pieces that can become embedded in the surrounding tissue.

## BONE AND JOINT INJURIES

A *dislocation* occurs when a bone is displaced from its proper position in a joint. This may result in the tearing and stretching of the ligaments. Reduction or return of the bone

UNIT 8 REPRESENTATIVE DISORDERS OF BONES AND JOINTS **69**

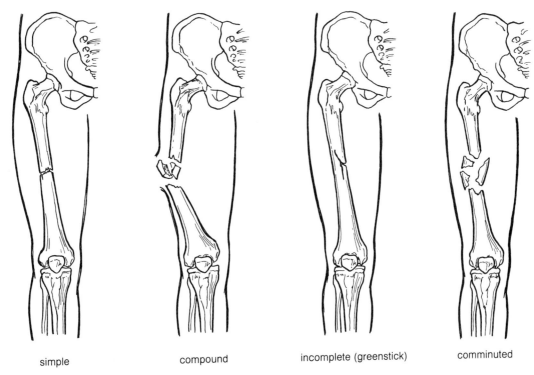

simple    compound    incomplete (greenstick)    comminuted

Figure 8-1 Types of fractures

to its proper position is necessary, along with rest to allow the ligaments to heal.

A *sprain* is an injury to a joint caused by any sudden or unusual motion, such as "turning the ankle." The ligaments are either torn from their attachments to the bones or torn across, but the joint is not dislocated. A sprain is accompanied by rapid swelling and acute pain in the area. Treatment consists of supporting the joint until the ligaments heal. This is usually done with adhesive strapping or an Ace bandage.

*Ankylosis* occurs when a joint becomes completely immobile because the bones have fused solid.

*Arthritis* is an inflammatory condition of one or more joints, accompanied by pain and often by changes in bone position. There are several types, the most common being rheumatoid arthritis, osteoarthritis, and gouty arthritis:

- *Rheumatoid arthritis* is a chronic systemic disease affecting the connective tissues and joints. There is acute inflammation of the connective tissue, thickening of the synovial membrane, and ankylosis of joints. The joints are badly swollen and painful. The pain, in turn, causes muscle spasms which may lead to deformities in the joints. In addition, the cartilage that separates the joints will degenerate, and hard calcium fills the spaces. When the joints become stiff and immobile, muscles attached to these joints slowly atrophy. This disease affects approximately three times more women than men. Its cause is unknown, although everything from emotional factors to endocrine and metabolic disorders has been cited.

- *Osteoarthritis* is a degenerative joint disease where the cartilage softens and degenerates, stimulating the formation of new bone at the joints. The joints affected are those receiving a great deal of wear and tear, particularly the lower extremity joints and the spine. Individuals over 45 years of age are most commonly affected.
- *Gouty arthritis* ("gout") is caused by a faulty uric acid metabolism. The level of uric acid is elevated in the bloodstream. Eventually uric acid crystals deposit in the joints, especially the metatarsophalangeal joint of the big toe.

A *bunion* is the swelling of the bursa of the foot, usually of the metatarsophalangeal joint in the big toe. (A bursa is a small sac between parts that move on each other.) Bunions result from poorly fitting shoes, poor walking posture, or a genetic tendency. The big toe then becomes adducted.

*Rickets* is a disease of the bones which is caused by a lack of vitamin D. Portions of the bones are soft, due to lack of calcification. The soft bones bend, causing such deformities as bowlegs and pigeon breast. The disease may be prevented by providing a growing child with sufficient quantities of calcium, vitamin D, and exposure to sunshine.

*Clubfoot* (talipes) is a congenital (existing at birth) malformation. It may involve one or both feet. The deformity may take one of several forms: the body weight may rest on the heel or ball of the foot only, or the inner or outer side of the sole may touch the ground.

*Microcephalus* is a congenital or hereditary (inborn) condition in which there is a marked diminution in the size of the cranium due to early ossification of the sutures of the skull. This is usually accompanied by arrested mental development.

*Spina bifida* is a congenital condition in which the vertebral column, which contains the spinal cord, does not develop completely and unite properly. This disorder usually affects the lumbar and sacral regions. Frequently, the contents of the spinal cord protrude out of the spinal cavity.

*Osteoporosis,* or softening of the bones, is caused by a deficiency in male or female hormones. The bones fracture easily because of brittleness. Replacement of hormones and increased mineral intake may slow the process.

*Bursitis* is the acute inflammation of the synovial bursa, which cushions a joint during motion. It can be caused by local or systemic inflammation or excess tension. Calcium deposits may form within the elbow, knee, and shoulder joints. This accumulation of calcium impedes movement of the joint.

*Webbed and extra fingers and toes* are congenital conditions which must be corrected by early surgery.

*Scoliosis* is a side-to-side or lateral curvature of the spine.

*Kyphosis* ("hunchback") is a humped curvature in the thoracic area of the spine.

*Lordosis* ("swayback") is an exaggerated inward curvature in the lumbar region of the spine just above the sacrum.

*Subluxation* is the most common problem associated with neck injuries. In this condition, a vertebra is displaced from its normal position, or normal range of movement, without having been completely dislocated. A subluxated neck vertebra can result from an automobile injury, especially whiplash, or other sudden, unexpected body movements. A sharp blow to the chin, face or head, even an uncontrolled coughing fit or a violent sneeze can subluxate a vertebra, figure 8-2.

## OTHER MEDICALLY RELATED DISORDERS

Bone disorders also occur with *tuberculosis, osteomyelitis,* and benign or malignant *tumors,* to name a few. *Rheumatic fever* causes the inflammation of connective tissue, notably in the heart and blood vessels, and around joints, synovial tissues and tendons. A common clinical symptom is arthritis.

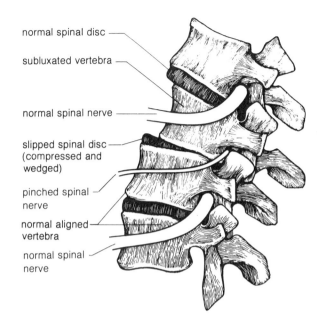

Figure 8-2 Subluxation of a vertebra. The second vertebra is out of alignment. Note the effect on the spinal disc and nerve. (Courtesy of CPR, Teaneck, New Jersey)

## Further Study and Discussion

- Make drawings of the different types of fractures as they would look if the leg were broken.

## Assignment

1. What is the difference between a simple and compound fracture?

2. Explain why a child with rickets would have bowlegs and other bone deformities. How can this disease be prevented?

3. What is a sprain? How can a sprain be treated?

4. Why is any form of arthritis considered to be a degenerative process?

5. State three congenital bone disorders and name the part of the body affected by each.

6. What is bursitis?

# SELF-EVALUATION

## Section 2 THE BODY FRAMEWORK

A. Match each term in column I with its correct description in column II.

| Column I | Column II |
|---|---|
| _____ 1. atlas | a. also called innominate bones |
| _____ 2. axis | b. bone of the upper arm |
| _____ 3. femur | c. fibrous band which joins bone to bone |
| _____ 4. frontal bone | d. contents of medullary canal |
| _____ 5. humerus | e. necessary for growth and repair of bone tissue |
| _____ 6. ligament | f. name for thigh bone |
| _____ 7. mandible | g. vertebra on which skull rests |
| _____ 8. pelvic girdle | h. name for forehead |
| _____ 9. periosteum | i. name of lower jaw |
| _____ 10. red marrow of bone | j. vertebra on which head rotates |
| _____ 11. tendon | k. makes red blood cells |
| _____ 12. yellow marrow of bone | l. joins a muscle to a bone |
| | m. name of upper jaw |

B. List five functions of the skeletal system.

C. State the difference between a dislocation and a sprain.

D. Explain the location and functions of the periosteum and Haversian canals.

73

# Section 3
# Body Movement

# Unit 9
# INTRODUCTION TO THE MUSCULAR SYSTEM

## KEY WORDS

cardiac muscle
locomotion
multinucleate
myoglobin
red muscle
skeletal (striated) muscle
smooth (visceral) muscle
sphincter
syncytium
white muscle

## OBJECTIVES

- Describe the functions of muscles
- Describe each of the three muscle types
- Define the Key Words related to this unit of study

The ability to move is an essential activity of the living human body which is made possible by the unique function of contractility in muscles.

Muscles comprise a large part of the human body; nearly half our body weight comes from muscle tissue. If you weigh 140 pounds, about 60 pounds of it comes from the muscles attached to your bones. Collectively, there are over 600 different muscles in the human body. These muscles allow us to move our bodies from place to place (locomotion) as well as move individual parts of the body. They help the body stay erect and determine posture, while producing most of the body's heat. Muscles also participate in the less obvious movements of the internal organs. In addition to playing a role in movement, muscles give the body its characteristic form. The skeleton determines the overall body shape, but the muscles that drape the skeleton produce the contours we perceive as beautiful and graceful.

## TYPES OF MUSCLES

All body movements are determined by three principle types of muscles. They are skeletal, smooth, and cardiac muscle. These muscles are also described as striated, spindle-shaped, and nonstriated because of the way their cells look under the ordinary compound light microscope.

*Skeletal muscles* are attached to the bone of the skeleton. They have cross bandings (striations) of alternating light and dark bands running perpendicular to the length of the muscle, figure 9-1. Because of this appearance, they are called *striped* or *striated* muscle; also voluntary muscle, because they contain nerves under voluntary control. Skeletal muscle is composed of bundles of muscle cells. Each cell is multinucleate (containing many nuclei). This special type of cell is called a *syncytium*.

The fleshy body parts are made of skeletal muscles. They provide movement to the limbs, but contract quickly, fatigue easily and lack the ability to remain contracted for

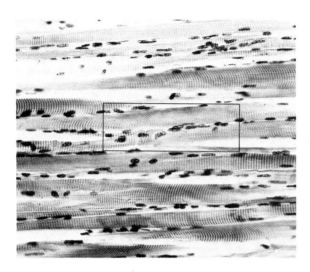

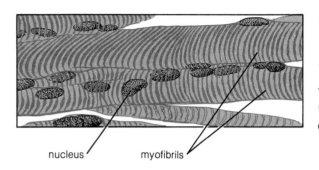

Figure 9-1 Voluntary or striated (skeletal) muscle cells (Photograph courtesy of Armed Forces Institute of Pathology, negative 72-13786)

prolonged periods. Blinking the eye, talking, breathing, dancing, eating and writing are all produced by the motion of these muscles.

There are two kinds of skeletal muscle. We can readily see this when we examine chicken that has been cooked. There are two kinds of meat: the so-called "dark meat" and "light meat," or *red muscle* and *white muscle*. The difference in color is due to the presence of the red pigment *myoglobin*. Myoglobin is richer in red muscle and turns brown when heated, resulting in dark meat. Myoglobin is a protein, which can bind to oxygen molecules. It is similar to the hemoglobin found in red blood cells. When oxygen is plentiful, skeletal muscle binds and stores myoglobin until needed, as during vigorous muscle contraction.

*Smooth (visceral) muscle* cells are small and spindle-shaped. There is only one nucleus, located at the center of the cell. They are called smooth muscles because they are unmarked by any distinctive striations. Unattached to bones, they act slowly, do not tire easily, and can remain contracted for a long time, figure 9-2.

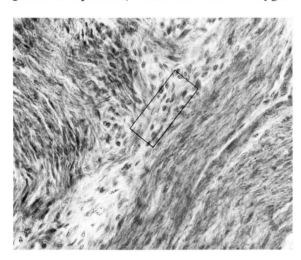

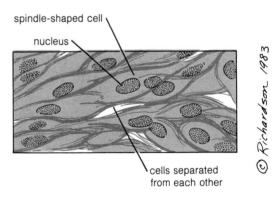

Figure 9-2 Involuntary or smooth muscle cells (Photograph courtesy of Armed Forces Institute of Pathology, negative 71-9163)

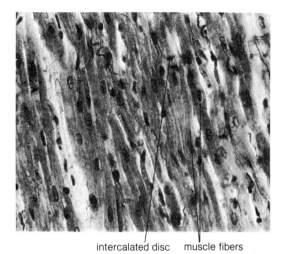 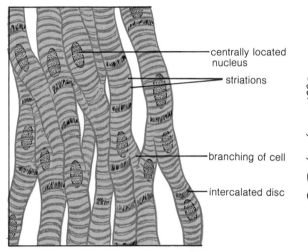

Figure 9-3 Cardiac muscle cells (Photograph reprinted, by permission, from Joan G. Creager, *Human Anatomy and Physiology,* 97)

Smooth muscles are not under conscious control; for this reason they are also called involuntary muscles. Their actions are controlled by the autonomic (automatic) nervous system. Smooth muscles are found in the walls of the internal organs, including the stomach, intestines, uterus, and blood vessels. Thus they help push food along the length of the alimentary canal, contract the uterus during labor and childbirth, and control the diameter of the blood vessels as the blood circulates throughout the body.

*Cardiac muscle* is found only in the heart. Cardiac muscle cells are striated and branched, and they are involuntary, figure 9-3. Healthy cardiac muscle contracts rapidly and is very strong. It is well suited to a lifetime function of pumping blood throughout the body. When the heart beats normally, it holds a rhythm of about 72 beats per minute. However, the activity of various nerves leading to the heart can increase or decrease its rate. Cardiac muscle requires a continuous supply of oxygen to function. Should its oxygen supply be cut off for as little as thirty seconds, it would stop beating.

*Sphincter muscles* are special circular muscles in the walls of the anus and the urethra. They open and close to control the passage of substances.

## Further Study and Discussion

- Observe prepared slides of muscle tissue or bring several types of uncooked meat to class (tripe, steak and heart). Using dissecting needles, place a few fibers in a drop of water on a slide. Observe each under the compound microscope. Notice the different kinds of fibers. Draw and label these fibers. Describe how they differ in appearance.

- Make up a chart listing the 3 muscle types and give several examples of each type.

## Assignment

A. Place the most correct answer in the space or spaces provided.

1. Muscles help to keep the body erect and therefore determine our _____.

2. Muscles produce most of the _____ that is generated in the body.

3. The action of muscles upon bones is responsible for movements of our _____.

4. A specialized muscle, the heart, is responsible for _____ throughout the body.

5. Muscle tissue helps get carbon dioxide out and oxygen into the body through the _____.

6. Three principal types of muscle tissue are the _____, _____, and the _____.

7. Cardiac muscle is an _____ muscle.

B. Answer the following questions.

1. What is myoglobin?




2. What is meant by "red muscle"?

3. List three characteristics of smooth muscle.

4. What part of the nervous system controls involuntary muscle actions?

5. Name four actions that are controlled by skeletal muscles.

# Unit 10
# ATTACHMENT OF MUSCLES

## KEY WORDS

abduction
adduction
antagonist
belly of muscle
biceps
contraction
depressor
dilator muscle
elastic
extension
extensor
flexor
insertion
levator
muscle fatigue
muscle tone
origin
pronation
sphincter muscle
supination
triceps

## OBJECTIVES

- List the characteristics of muscles
- Describe how pairs of muscles work together
- Describe how muscles are attached
- Describe how muscles are ready for action
- Define the Key Words relating to this unit of study

All muscles, whether they are skeletal, smooth or cardiac, have three characteristics in common. One is *contractibility,* a quality possessed by no other body tissue. When a muscle shortens or contracts, it reduces the distance between the parts of its contents, or the space it surrounds. The contraction of skeletal muscles which connect a pair of bones brings the attachment points closer together. This causes the bone to move. When cardiac muscles contract, they reduce the area in the heart chambers, pumping blood from the heart into the blood vessels. Likewise, smooth muscles surround blood vessels and the intestines, causing the diameter of these tubes to decrease upon contraction.

Another property of muscles is *extensibility* (the ability to be stretched). When we bend our forearm, the muscles on the back of it are extended or stretched. Finally, muscles exhibit *elasticity* (ability of a muscle to return to its original length when relaxing). Collectively, these three properties of muscles — contractility, extensibility and elasticity — produce a veritable mechanical device capable of complex, intricate movements.

## ANTAGONISTIC MUSCLE PARTS

There are over 600 different muscles in the body. For any of these muscles to produce movement in any part of the body, it must be able to exert its force upon a movable object. In other words, muscles must be attached to bones for leverage in order to have something to pull against.

Muscles are attached to the bones of the skeleton by nonelastic cords called tendons. Bones are connected by joints. Skeletal muscles are attached in such a way as to bridge these joints. So, when a skeletal muscle contracts, the bone to which it is attached will move.

Muscles are attached at both ends to bones, cartilage, ligaments, tendons, skin and sometimes to each other. The *origin* is the part of a skeletal muscle that is attached to a fixed structure or bone; it moves least during muscle contraction. The *insertion* is the other end, attached to a movable part; it is the part that moves most during a muscle contraction. The *belly* is the central body of the muscle, figure 10-1.

The muscles of the body are arranged in pairs. One produces movement in a single direction, the other does so in the opposite direction. This arrangement of muscles with opposite actions is known as an *antagonist* pair.

By example, upper arm muscles are arranged in antagonist pairs, figure 10-1. The muscle located on the front part of the upper arm is the *biceps*. One end of the biceps is attached to the scapula and humerus (its origin). When the biceps contracts, these two bones remain stationary. The opposite end of the biceps is attached to the radius of the lower arm (its insertion); this bone moves upon contraction of the biceps.

The muscle on the back of the upper arm is the *triceps*. Try this simple demonstration: Bend your elbow. With your other hand, feel the contraction of the belly of the biceps. At the same time, stretch your fingers out (around the arm) to touch your triceps; it will be in a relaxed state. Now extend your forearm; feel the simultaneous contraction of the triceps and relaxation of the biceps. Now bend the forearm halfway and contract the biceps and triceps. They cannot move, since both sets of muscles are contracting at the same time.

The biceps is a *flexor* muscle, because it flexes or bends a joint, in this case the elbow. The triceps is an *extensor* muscle, since it extends or straightens a joint. In addition to the elbow joint, antagonistic flexors and extensors are located at the ankle, knee, wrist

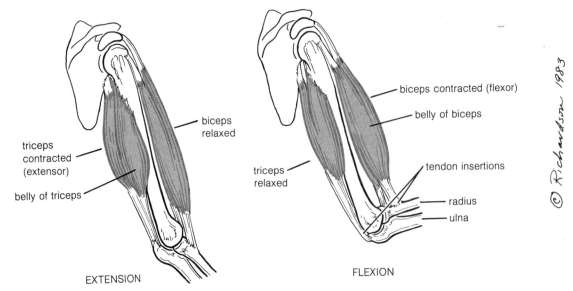

Figure 10-1 Coordination of antagonistic muscles

and several other areas. There are other types of antagonistic movements controlled by skeletal muscles. *Adduction* moves parts of the body toward the midline of the body; *abduction* moves parts of the body away from the midline. Levator and depressor muscles raise and lower body parts, as in raising and lowering the mandible when chewing or speaking. Then we have *pronation* and *supination:* pronation rotates the forearm so that the palm of the hand is turned towards the body, while supination turns the palm away. Finally, there are sphincter muscles and dilators — which decrease or increase openings — such as the muscles around the anus and the mouth.

Blood circulation, body tissues, and the liver supply oxygen and sugar from digested food, necessary for the function of the muscle cells. Muscles and liver store sugar and glycogen, which they can convert to glucose.

Skeletal muscle, in addition to helping us move, maintains our posture and produces heat. Living human beings must constantly maintain their body temperature within a rather narrow range (98.6°F or 37°C). As a result of catabolism, muscles produce heat and wastes. Since skeletal muscles make up close to half our total body weight, they generate much of the body's heat.

## MUSCLE FATIGUE

Muscle fatigue is caused by an accumulation of lactic acid in the muscles. During periods of vigorous exercise, the blood is unable to transport enough oxygen for the complete oxidation of glucose in the muscles. This causes the muscles to flex and contract anaerobically (without oxygen).

Aerobic oxidation is a chemical process whereby energy and pyruvic acid are released from sugar. The pyruvic acid eventually is converted to $CO_2$ and waste, which is then excreted from the body. However, in anaerobic (insufficient oxygen) oxidation of muscles, the pyruvic acid is converted into lactic acid. The lactic acid normally leaves the muscle, passing into the bloodstream. But if vigorous exercise continues, the lactic acid level in the blood rises sharply. In such cases, lactic acid accumulates within the muscle. This impedes muscular contraction, causing muscle fatigue and cramps.

## MUSCLE TONE

In order to function rapidly and well, muscles should always be slightly contracted and ready to pull. This is *muscle tone.* People in good health have firm muscles which are always ready to work. Muscle tone can be achieved through proper nutrition and regular exercise. Each muscle is in contact with the nervous system through the motor nerve which carries messages from the brain to the muscles and makes them always ready for action.

### Further Study and Discussion

- Ask your butcher for a chicken leg with the foot still intact. Cut open the skin and free the ends of the tendons. Pull the tendons in the front of the leg, then those behind it. What happened to the toes? Did the tendons stretch?
- Discuss what happens to muscles that are not used.

- Make arrangements for a physiotherapist to speak about maintaining muscle tone in the elderly patient.

## Assignment

A. Complete the following sentences.
   1. Muscles are arranged in _____, one muscle being _____ to the other and performing an action _____ to that of the other.
   2. A muscle which bends at the joint is called a _____.
   3. The muscle which bends a joint appears thicker and shorter than the one that straightens the joint, called an _____.
   4. Muscles may be attached to _____, _____, _____, _____, _____, and sometimes to _____.
   5. The end of the muscle which moves least during muscle contraction is the _____.
   6. The end of the muscle moving the most is the _____.
   7. The nourishment necessary for the work of the muscle cells is furnished by the _____ and carried by the _____.
   8. Muscles and the liver store _____ which may be converted to glucose.
   9. Muscle fatigue is caused by an accumulation of _____, a result of _____ _____ of stored sugar.
   10. Oxidation is the main _____ _____ by which _____ is released from the sugar in the cells.

B. Answer the following questions.
   1. List the three properties of muscles.

2. Explain what is meant by an antagonistic muscle pair.

3. Describe how a muscle becomes fatigued.

# Unit 11
# PRINCIPAL SKELETAL MUSCLES

## OBJECTIVES

- Locate the important skeletal body muscles
- Describe the function of these muscles
- Identify the muscles using the technical names

Table 11-1 lists the principal skeletal muscles, their locations and functions. The muscle groups as they appear on the human body can be studied in figure 11-1.

Table 11-1 Location and Function of Principal Skeletal Muscles

| MUSCLE | LOCATION | FUNCTION |
|---|---|---|
| Sternocleidomastoid | Neck | Moves head |
| Deltoid | Shoulder | Abducts upper arm |
| Biceps (brachii) | Upper arm | Flexes lower arm |
| Triceps (brachii) | Upper arm | Extends lower arm |
| Pectoralis major | Anterior chest | Flexes upper arm; helps adduct upper arm |
| Intercostals | Between ribs | Move ribs (assist in breathing) |
| Diaphragm | Between abdominal and chest cavities | Enlarges thorax (assists in breathing) |
| Serratus | Anterior chest | Moves shoulder |
| Rectus abdominus | Extends from ribs to pelvis | Compresses abdomen |
| Sartorius | Anterior thigh | Flexes and rotates thigh and leg |
| Rectus femoris | Anterior thigh | Flexes thigh; extends lower leg |
| Vastus lateralis | Anterior thigh | Extends leg |
| Tibialis anterior | Anterior leg | Flexes and elevates foot |
| Extensors | Ankle and foot | Move foot and toes |
| Extensors Flexors | Forearm | Move hand and fingers |
| Trapezius | Posterior chest and back | Moves shoulder; extends head |
| Latissimus dorsi | Posterior chest and back | Extends upper arm; helps adduct and rotate upper arm |
| Gluteus medius | Buttocks | Abducts and rotates thigh |
| Gluteus maximus | Buttocks | Extends thigh and rotates it outward |
| Biceps femoris | Posterior thigh | Flexes leg; extends thigh |
| Gastrocnemius | Posterior leg | Points toes; flexes lower leg |
| Sacrospinalis | Inserted in ribs and vertebrae | Extends spine; abducts and rotates trunk |

UNIT 11 PRINCIPAL SKELETAL MUSCLES 87

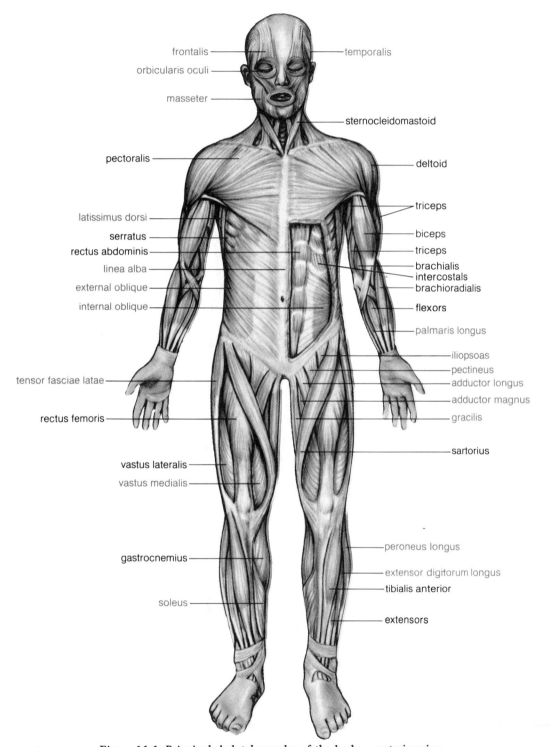

Figure 11-1 Principal skeletal muscles of the body — anterior view

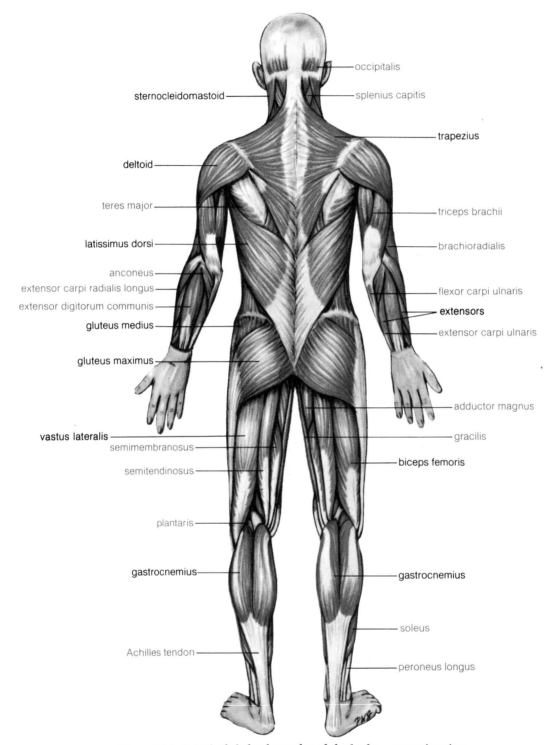

Figure 11-1 Principal skeletal muscles of the body — posterior view

## Further Study and Discussion
- Identify muscles which are used for giving injections.
- Identify the muscles which seem to work in pairs.
- Discuss the effect of massage on muscles.

## Assignment
Give the general function of the listed muscles.

1. Deltoid

2. Intercostals

3. Rectus Femoris

4. Gluteus Maximus

5. Triceps (brachii)

6. Pectoralis major

7. Sartorius

8. Extensors

9. Trapezius

10. Tibialis anterior

# Unit 12

# REPRESENTATIVE MUSCULO-SKELETAL DISORDERS

## KEY WORDS

abdominal hernia
flatfeet (talipes)
inguinal hernia
muscle atrophy
muscle fatigue
muscle
  hypertrophy
muscle spasms
muscular dystrophy
myasthenia gravis
poliomyelitis
rehabilitation
stiff neck
tetanus (lockjaw)

## OBJECTIVES

- Identify some common muscle disorders
- Describe how conditions and diseases may prevent proper muscle function
- Define the Key Words related to this unit of study

Muscular coordination is very important if a person is to perform his/her daily functions efficiently. Injuries and diseases which may affect muscles sometimes interfere with these functions. The retraining of injured or unused muscles is a type of *rehabilitation,* called therapeutic exercise.

*Muscle atrophy* can occur to muscles which are infrequently used; they shrink in size and lose muscle strength. An example is *poliomyelitis* (polio), where there is damage or paralysis of nerves carrying nerve messages to the muscles. The muscles are understimulated, and gradually waste away. Muscle atrophy due to nerve paralysis may reduce a muscle up to 25% of its normal size. The wasted muscle is replaced by non-contractile connective tissue. Muscle atrophy can also be caused by prolonged bedrest or the immobilization of a limb in a cast. Muscle atrophy can be minimized by direct electrical stimulation, massage, or special exercise.

*Flatfeet* ("fallen arches" or talipes) result from a weakening of the leg muscles that support the arch. The downward pressure on the foot eventually flattens out the arches. Muscle strength can be increased by exercise, massage, and electrical stimulation.

An *abdominal hernia,* or rupture, may occur in a weak place in the muscular abdominal wall. It is caused by bulging of the intestine through an opening in the wall of the abdominal cavity normally containing it. The inguinal hernia is the most frequent type of hernia. It appears in the groin area.

*Muscle hypertrophy* is a condition in which a muscle enlarges and grows stronger. It results from overworking or overexercising. This leads to an increase in the diameter or size of muscle cells, as opposed to an increase in the number of muscle cells. So a change in girth increases the total force of the muscle's contraction.

*Muscle fatigue* may occur from the temporary overuse of a muscle. Fatigue lessens the muscle's ability to perform work. (Review unit 10.)

*Stiff neck* may be due to an inflammation of the trapezius muscle. The rigidity is the result of unusual overuse of the muscle.

*Muscle spasms* are sudden and violent contractions caused by sudden overworking of the muscle or by poor circulation to the localized area.

*Tetanus (lockjaw)* is an infectious disease, usually fatal, characterized by continuous spasms of the voluntary muscles. It is caused by a toxin from a tetanus bacillus, *Clostridium tetani*, which can enter the body through any open wound (especially a puncture wound).

*Muscular dystrophy* is a chronic wasting disease of the muscles. It often appears during childhood and is thought to result from some genetic disturbance.

*Myasthenia gravis* leads to progressive muscular weakness and paralysis, sometimes even death. The cause is still unknown, but many researchers believe it may be due to a defect in the immune system, particularly the thyroid gland. In extreme cases, it can be fatal due to the paralysis of the respiratory muscles.

## Further Study and Discussion

- Visit a physical rehabilitation center. Report on the success of muscle retraining in this program.
- Discuss how neglectful patient care can lead to muscular atrophy.
- Invite a physical therapist to discuss rehabilitation with your class.

## Assignment

Match each term in column I with its correct description in column II.

| Column I | Column II |
|---|---|
| _____ 1. muscular atrophy | a. the retraining or rehabilitation of muscle use |
| _____ 2. muscular dystrophy | b. the temporary overuse of a muscle |
| _____ 3. paralysis | c. chronic wasting of the muscle tissue |
| _____ 4. stiff neck | d. sudden and violent muscle contraction |
| _____ 5. muscle fatigue | e. continuous spasm caused by a toxin |
| _____ 6. muscle hypertrophy | f. immobility caused by blocked nerve messages |
| _____ 7. muscle spasm | g. bulging of an organ through a muscular wall |
| _____ 8. hernia | h. major loss of muscle strength and size |
| _____ 9. therapeutic exercise | i. rigidity often caused from inflammation of the trapezius muscle |
| _____ 10. tetanus | j. muscle enlargement due to overworking of the muscle |
| | k. condition resulting from weak arch muscles |

# SELF-EVALUATION

## Section 3
## BODY MOVEMENT

A. Match each term in column I with its correct function in column II.

| Column I | Column II |
|---|---|
| ___ 1. biceps | a. extends upper arm |
| ___ 2. diaphragm | b. assists in breathing |
| ___ 3. gastrocnemius | c. flexes lower arm |
| ___ 4. gluteus medius | d. extends spinal column and moves trunk |
| ___ 5. latissimus dorsi | e. moves foot and leg |
| ___ 6. sacrospinalis | f. closes body openings |
| ___ 7. serratus | g. abducts and rotates thigh |
| ___ 8. sphincter | h. moves shoulder |
| | i. moves the head |

B. Define the following terms.

1. Flatfoot

2. Bursitis

3. Hernia, or rupture

4. Muscular atrophy

5. Paralysis

6. Tetanus

# Section 4 Transport of Food and Oxygen

# Unit 13

# INTRODUCTION TO THE CIRCULATORY SYSTEM

## KEY WORDS

acid
acid-base alkali
base
bicarbonate
closed circulatory system
phosphate
plasma
platelets
pulmonary circulation
systemic circulation

## OBJECTIVES

- Describe the function of the circulatory system
- List the components of the circulatory system
- Describe the two routes of blood circulation
- Define the Key Words relating to this unit of study

Blood is an essential life supportive fluid, transported throughout the body through a system of blood vessels. This is known as a *closed blood vessel system,* a major characteristic of vertebrates.

The circulatory system is the largest organ of the body. If one were to lay all of the blood vessels in a single human body end to end, they would stretch one fourth the way from earth to the moon, a distance of some 60,000 miles.[1]

## FUNCTIONS OF BLOOD

Blood performs numerous functions in helping the body to maintain a stable internal environment. This environment is essential to the functioning of the body's various activities.

Blood is a liquid tissue, composed of a fluid component (plasma), and a solid component (blood cells). Blood cells include red blood cells, white blood cells, and platelets. The red blood cells convey oxygen to the cells for oxidation, and carbon dioxide away for excretion. White blood cells protect against disease by engulfing and digesting bacteria and foreign matter which has invaded the body. The platelets help the blood to clot whenever internal or external bleeding occurs.

Plasma suspends blood cells and transports them throughout the body. In addition to bringing dissolved nutrients to the cells, plasma also carries metabolic waste products from the cells to the various excretory organs, where they are excreted or converted into compounds useful for other purposes. Endocrine glands, which synthesize chemical compounds called hormones, secrete these hormones into the plasma. The plasma then

---
1. I. Sherman and V. Sherman, *Biology: A Human Approach* (New York: Oxford University Press, 1979).

circulates them to various body parts in order to help regulate bodily functions.

Blood also helps the body to maintain its water content and body temperature. It is essential that the body's temperature does not rise too far above 98.6°F or 37°C. High body temperature will disrupt important chemical reactions in the cells. Generally, highly active muscle tissue creates a good deal of heat that can raise the body's temperature. To counter this effect, blood then circulates more quickly through the tissue to diffuse the heat. This excess heat is subsequently given off over the body's surface through the skin.

Finally, blood helps to maintain the body's internal acid-base balance. Chemicals in the bloodstream, known as *bicarbonates* and *phosphates,* neutralize small amounts of *acids* or *alkalis* (basic compounds).

## COMPONENTS OF THE CIRCULATORY SYSTEM

As mentioned earlier, the blood itself contains plasma, red blood cells, white blood cells, and platelets. They are transported within a closed circulatory system. In such a system, blood does not, under normal circumstances, leave the blood vessels to flow among the tissues. Rather, it remains inside the blood vessels and is transported throughout the body to form a closed circuit of blood. This blood circuit is composed of arteries, arterioles, veins, venules, and capillaries. Blood is pumped through the blood vessels by the action of a muscular pump, which we know as the heart.

The circulatory system also includes the lymphatic system. This consists of the lymph and tissue fluid derived from the blood and the lymphatic vessels, which return the lymph to the blood. The spleen is considered a part of the circulatory system. It provides a reservoir for blood and is active in destroying microorganisms in the blood.

## MAJOR BLOOD CIRCUITS

Blood leaves the heart through arteries and returns by veins. The blood uses two circulation routes:

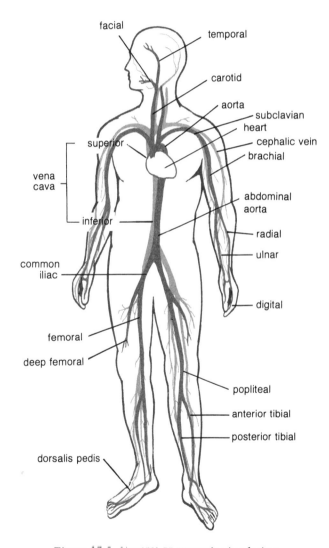

Figure 13-1 General or systemic circulation

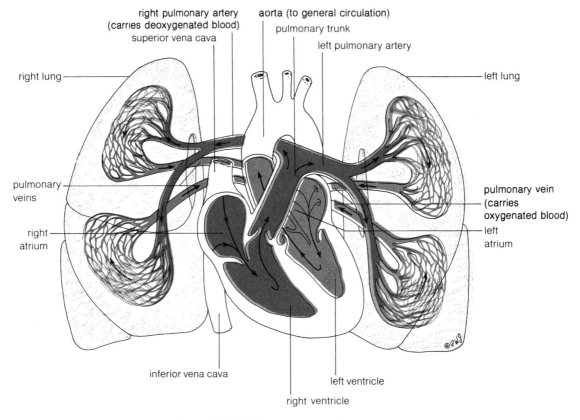

**Figure 13-2 Pulmonary circulation**

1. The *general* (or *systemic*) *circulation* carries blood throughout the body, figure 13-1.

2. The *pulmonary circulation* carries blood from the heart to the lungs and back, figure 13-2.

## Further Study and Discussion

- If available, observe blood cells circulating through capillaries in the web of a frog's foot, or in the tail of a goldfish. Note the direction of the blood flow.
- Discuss how blood from the general circulation is oxygenated by blood from the pulmonary circulation.

## Assignment

A. Briefly answer the following questions.

1. Describe the chief functions of the circulatory system.
2. What is the name of the large vein which returns blood to the heart from the general circulation?
3. What is meant by a closed circulatory system?
4. What is the composition of blood? List the components of the circulatory system.

B. Match each term in column I with its description or function in column II.

| Column I | Column II |
| --- | --- |
| _____ 1. pulmonary artery | a. vein which carries freshly oxygenated blood from the lung to the heart |
| _____ 2. lymphatic system | b. circulation route which carries blood to and from the heart and lungs |
| _____ 3. pulmonary vein | c. organ which provides a reservoir for blood, destroys microorganisms in the blood, and removes worn out blood cells |
| _____ 4. spleen | d. artery which carries deoxygenated blood from the heart to the lung |
| _____ 5. pulmonary circulation | e. system which consists of lymph and tissue fluid derived from the blood |
| _____ 6. left ventricle | f. blood from the pulmonary vein which re-enters the heart through the right atrium |
| _____ 7. general circulation | g. artery which carries blood with nourishment, oxygen and other materials from the heart to all parts of the body |
| _____ 8. right ventricle | h. ventricle from which the aorta receives blood |
| _____ 9. aorta | i. circulation which carries blood throughout the body |
| | j. ventricle from which the pulmonary artery leaves the heart |

# Unit 14
# THE HEART

## KEY WORDS

aortic semilunar valve
apex
atrioventricular bundle (bundle of His)
atrioventricular node
atrioventricular valve
atrium
auricle
bicuspid (mitral) valve
cardiac arrest
cardiac arrhythmia
cardiopulmonary resuscitation (CPR)
deoxygenate
endocardium
fibrous pericardium
foramen ovale
myocardium
oxygenate
pericardial fluid
pericardium
pulmonary semilunar valve
purkinje network
septum
serous pericardium
sinoatrial node (pacemaker)
stethoscope
tricuspid valve
ventricle

## OBJECTIVES

- Describe the structure of the heart
- Describe the function of various parts of the heart
- Locate and identify various parts of the heart
- Define Key Words relating to this unit of study

The blood's circulatory system, like other systems of the body, is extremely efficient. The main organ responsible for this efficiency is the heart, a tough, simply-constructed muscle about the size of a closed fist.

The adult human heart is about 5 inches long and 3.5 inches wide, weighing less than a pound (12-13 oz.). The importance of a healthy, well-functioning heart is obvious: to circulate life-sustaining blood throughout the body. When the heart stops beating, life stops as well! To explain further, if the blood flow to the brain ceases for 5 seconds or more, the subject loses consciousness. After 15-20 seconds, the muscles twitch convulsively; after 9 minutes without blood flow, the brain cells are irreversibly damaged.

The heart is located in the mediastinal cavity. This places the heart between the lungs, behind the sternum, in front of the thoracic vertebrae and above the diaphragm. Although the heart is centrally located, its axis of symmetry is not along the midline. The heart's apex (conical tip) lies on the diaphragm and points to the left of the body. It is at the apex where the heartbeat is most easily felt and heard through the *stethoscope*.

Try this simple demonstration: place the disk or bowl of a stethoscope over the heart's apex. This is the area between the fifth and sixth ribs, along an imaginary line extending from the middle of the left clavicle. Since the heartbeat is felt and heard so easily at the apex, this gives rise to the popular but incorrect notion that the heart is located on the left side of the body.

Knowledge of the correct position of the heart can make all the difference in the treatment of *cardiac arrest*. Using the heel of the hand, one applies a series of sharp, forceful, pushing motions to the lower sternum of the victim so as to restart the heartbeat. During such a medical emergency, the combination of manual heart compression and artificial respiration can save a life. This life-saving technique is known as *cardiopulmonary resuscitation (CPR)* and should be performed only by those specifically trained in CPR.

## STRUCTURE OF THE HEART

The heart is a hollow, muscular, double pump which circulates the blood through the blood vessels to all parts of the body. At rest, the heart pumps two ounces of blood with each beat, five quarts per minute, seventy-five gallons per hour.

Surrounding the heart is a double layer of fibrous tissue called the *pericardium*. Between these two pericardial layers is a space filled with a lubricating fluid called *pericardial fluid*. This fluid prevents the two layers from rubbing against each other and creating friction. The thin inner layer covering the heart is the *serous pericardium*. The tough outer membrane is the *fibrous pericardium*.

Cardiac muscle tissue, or *myocardium*, makes up the wall of the heart. On the inner lining lies a smooth tissue called the *endocardium*. The endocardium covers the heart valves and lines the blood vessels providing smooth transit for the flowing blood.

A frontal view of the human heart reveals a thick, muscular wall separating it into a right half and a left half. This partition, known as the *septum*, completely separates the blood in the one half from that in the other half, see color plate 5.

In the human fetus, however, there is an opening in the septum, *foramen ovale*, which connects the two sides of the heart. This condition exists uniquely in the human fetus and in the newborn infant. The opening allows blood to flow freely from the left side to the right side without passing through the lungs for oxygen. Only after birth do the lungs begin to function as an oxygen source, whereupon the foramen ovale closes, forming the solid septum. After this, the right side of the heart carries only *deoxygenated*\* (without oxygen) blood and the left side carries only oxygenated (with oxygen) blood.

The human heart is separated into right and left halves by the septum. In turn, each half is divided into two parts, thus creating four chambers. The two upper chambers are called the *right atrium* and the *left atrium* (pl. atria). The atrium is also referred to as the *auricle*. The lower chambers are the *right ventricle* and the *left ventricle*.

The heart has four *atrioventricular valves* which permit the blood to flow in one direction only. These valves keep the blood from flowing backwards into the chambers.

- The *tricuspid valve* is positioned between the right atrium and the right ventricle. Its name comes from the fact that there are three points, or cusps, of attachment (on the floor of the right ventricle). It allows blood to flow from the right atrium into the right ventricle, but not in the opposite direction.

- The *bicuspid (mitral) valve* is located between the left atrium and the left ventricle. Blood flows from the left atrium into the left ventricle, while backflow from the left ventricle to the left atrium is prevented.

- The *pulmonary semilunar valve* is found at the orifice (opening) of the pulmonary

---
\**Note:* Venous blood is said to be deoxygenated, meaning that it contains a large amount of carbon dioxide. However, a small amount of oxygen is present.

artery. It lets blood travel from the right ventricle into the pulmonary artery, and then into the lungs.

- The *aortic semilunar valve* is at the orifice of the aorta. This valve permits the blood to pass from the left ventricle into the aorta, but not backwards into the left ventricle.

# CONTROL OF HEART CONTRACTIONS

A heart completely removed from the body will continue to beat rhythmically if it is supplied with a proper nutrient media. This shows that heartbeat generates in the heart muscle itself. It is therefore said to be *myogenic* (myo – muscle; gennan – to produce; Gk.). The myocardium contracts rhythmically in order to perform its duty as a forceful pump. See color plate 6.

Control of heart muscle contractions is found within a group of nerve cells located at the opening of the superior vena cava into the right atrium. These cells are known as the *sino-atrial (S-A) node,* or *pacemaker.* The S-A node can only be identified by microscopic examination of the cardiac tissue. It sends out an electrical impulse which spreads out over the atria, making them contract simultaneously. This causes blood to flow downward from the upper atrial chamber to the atrioventricular openings. The electrical impulse eventually reaches the atrioventricular (A-V) node, which is another nerve cell group located between the atria and ventricle.

From the A-V node, the electrical impulse is carried to nerve fibers in the septum. These nerve fibers are known as the *atrioventricular bundle* or the *bundle of His.* It divides into a right and left branch; each branch then subdivides into a fine network of branches spreading throughout the ventricles, called the *Purkinje network.* The electrical impulse shoots along the Purkinje fibers, until it reaches the heart's apex.

The combined action of the S-A and A-V nodes is instrumental in the cardiac cycle. The cardiac cycle comprises one complete heartbeat, with both atrial and ventricular contractions.

1. The S-A node stimulates the contraction of both atria. Blood flows from the atria into the ventricles through the open tricuspid and mitral valves. At the same time, the ventricles are relaxed, allowing them to fill with the blood. At this point, since the semilunar valves are closed, the blood cannot enter the pulmonary artery or aorta.

2. The A-V node stimulates the contraction of both ventricles so that the blood in the ventricles is pumped into the pulmonary artery and the aorta through the semilunar valves which are now open. At this point the atria are relaxed and the tricuspid and mitral valves closed.

3. The ventricles relax; the semilunar valves are closed to prevent the blood flowing back into the ventricles. The cycle begins again with the signal from the S-A node.

## Further Study and Discussion

- If laboratory facilities are available, obtain a calf or sheep heart to examine. Notice the pericardium which is the tissue-like covering around the heart. Locate the atriums, ventricles and identify the four valves.
- On the following diagram locate and label the various structures of the heart. Also include valves, vessels, and nodes.

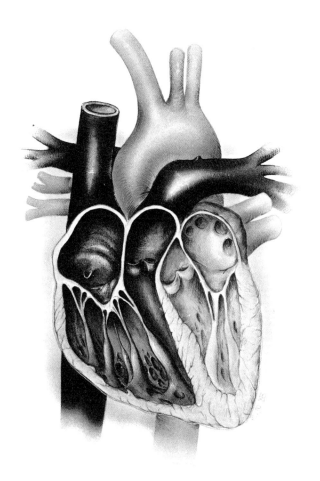

## Assignment

A. Briefly answer the following questions.

1. What is the pericardium? Describe its function.

2. Describe the endocardium and list its functions.

3. Name the four chambers of the heart.

4. Name each valve, then locate its position by identifying the structure or area on both sides of the valve.

5. What function does a valve perform?

B. Complete the following statements.
   1. The heart or cardiac muscle is called the _____.
   2. The _____ is the smooth lining of the heart.

3. The structure known as the _____ stimulates the contraction of the ventricles.

4. The ventricle with the thickest muscular structure is on the _____ side of the heart.

5. The _____ is also called the pacemaker of the heart.

# Unit 15

# FUNCTION AND PATH OF GENERAL CIRCULATION

## KEY WORDS

aorta
aortic arch
arteriole
bifurcate
brachiocephalic artery
coronary artery
coronary circulation
coronary sinus
coronary vein
hepatic portal vein
hepatic vein
inferior vena cava
left common carotid artery
left subclavian artery
portal circulation
portal vein
pulmonary artery
pulmonary vein
renal artery
renal circulation
renal vein
superior vena cava
systemic circulation
venule

## OBJECTIVES

- Trace the path of the general circulation
- Describe the four main functions of general circulation
- Name some specialized circulatory systems
- Define the Key Words relative to this unit of study

The function of the general (systemic) circulation is fourfold: it circulates nutrients, oxygen, water, and secretions to the tissues and back to the heart; it carries products such as carbon dioxide and other dissolved wastes away from the tissues; it helps equalize body temperature; it aids in protecting the body from harmful bacteria.

## THE GENERAL CIRCULATION

Let us trace a drop of blood through the human circulatory system. The blood can return to the heart from the arms or legs. Blood flowing from the upper part of the body (head, neck, and arms) returns to the right side of the heart via the *superior vena cava*. The superior vena cava is one of the two largest veins in the body. Blood from the lower part of the body (legs and trunks) enters the heart through the *inferior vena cava* (the other of the largest veins).

From both these veins, blood enters the right atrium, which then contracts, forcing the blood through the tricuspid valve into the right ventricle. This chamber contains deoxygenated blood from the tissues. Deoxygenated blood contains very little oxygen but a good deal of carbon dioxide.

The right ventricle then contracts to push the deoxygenated blood through the *pulmonary semilunar valve* into the pulmonary artery. The pulmonary artery bifurcates (divides in two). It branches into the *right pulmonary artery,* bringing deoxygenated blood to the right lung, and into the *left pulmonary artery,* bringing blood to the left lung.

Inside the lungs, the pulmonary arteries branch into countless small arteries called *arterioles.* The arterioles connect to dense beds of capillaries lying in the alveoli lung tissue. Here, gaseous exchange takes place: carbon dioxide leaves the red blood cells and is discharged into the air in the alveoli, to be excreted from the lungs. Oxygen, in turn, combines with hemoglobin in the red blood cells. From these capillaries the blood travels into small veins or *venules.*

Venules from the right lung eventually form two larger *pulmonary veins,* known as the *right pulmonary veins.* These veins carry oxygenated blood from the right lung back to the heart and into the left atrium.

From the left lung, blood also enters the left atrium via the two *left pulmonary veins.* Then the left atrium contracts, sending the blood through the bicuspid, or mitral valve, into the left ventricle. This chamber, then, acts as a pump for newly oxygenated blood. When the left ventricle contracts, it sends oxygenated blood through the aortic semilunar valve, then into the *aorta.*

The aorta is the largest artery in the body. As the aorta emerges (ascending aorta) from the anterior (upper) portion of the heart, it forms an arch. This arch is known as the *aortic arch.* Three branches come from this arch: the *brachiocephalic,* the *left common carotid,* and the *left subclavian* arteries, see color plate 7A. These arteries and their branches carry blood to the arms, neck and head.

From the aortic arch, the aorta descends along the mid-dorsal wall of the thorax and abdomen. Many arteries branch off from their descending aorta, carrying oxygenated blood throughout the body. The first branch from the descending aorta is the *coronary artery.* It carries blood to the heart's muscular wall.

As the descending aorta proceeds posteriorly, it sends off additional branches to the body wall, stomach, intestines, liver, pancreas, spleen, kidneys, reproductive organs, urinary bladder, legs, and so forth. Each of these arteries subdivides into still smaller arteries, then into arterioles, and finally into numerous capillaries embedded in the tissues. This is where hormones, nutrients, oxygen and other materials are transferred from the blood into the tissues.

In turn, metabolic waste products, such as carbon dioxide and nitrogenous wastes, are picked up by the blood. Hormones secreted by specialized tissues, and nutrients from the small intestines and liver, are also absorbed by the blood. Blood then runs from the capillaries first into tiny veins, through increasingly larger veins, and finally into one (or more) of the veins which exit from the organ. Eventually it empties into one of two largest veins in the body, see color plate 7B.

Deoxygenated venous blood, returning from the lower parts of the body, empties into the inferior vena cava. Venous blood from the upper body parts (arms, neck and head) passes into the superior vena cava. Both the inferior and superior vena cavae empty their deoxygenated blood into the right atrium.

## Coronary Circulation

Some circulatory routes within the general circulation are called by special names. The *coronary circulation,* which brings oxygenated blood to the heart, is a part of the general circulation. It has two branches, the *left* and *right coronary arteries.* These branches come off from the aorta just above the heart. The branches encircle the heart muscle, with many tiny branches going to all parts of the heart muscle. Blood returns to the right atrium by a pocket or trough in the wall of the right atrium, the *coronary sinus,* into which the *coronary veins* empty, see color plates 8 and 9.

## Renal Circulation

The *renal circulation* is that part of the general circulation which carries blood from the aorta through the renal artery to the kidneys and then through the renal vein back into the heart by way of the inferior vena cava. This renal vein carries limited amounts of cell waste since many of them have been removed from the blood in the kidneys.

## Portal Circulation

The portal circulation is a branch of the general circulation. Veins from the pancreas, stomach, small intestine, colon and spleen empty their blood into the portal vein, in the liver.

It is very important that venous blood makes a detour through the liver before returning to the heart. After meals, blood reaching the liver contains a higher than normal concentration of glucose. The liver removes the excess glucose, converting it to storage glycogen. In the event of vigorous exercise, work or prolonged periods without nourishment, glycogen reserves will be changed back into glucose for energy. This detour insures that the blood's glucose concentration is kept within a relatively narrow range.

The liver also detoxifies (neutralizes) drugs and toxins, and degrades hormones no longer useful to the body. The liver also removes worn-out red blood cells, or converts them into bile, and converts excess amino acids to urea.

Deoxygenated venous blood leaves the liver through the *hepatic portal vein,* which carries it to the inferior vena cava. From the inferior vena cava, blood enters the right atrium, see color plate 10.

## Assignment

A. Briefly answer the following questions.

1. Describe the chief functions of the general circulatory system.

2. Name the special circulatory system which is responsible for each of the following functions.
   a. Carries water to the kidneys

   b. Nourishes the heart

   c. Serves the digestive organs

B. Complete the following statements.
   1. The blood leaves the heart from the _____ ventricle through the _____, the largest artery in the body.
   2. The deoxygenated blood re-enters the heart through the _____ and into the _____ atrium.

C. Match each term in column I with its correct description in column II.

| | Column I | Column II |
|---|---|---|
| _____ | 1. coronary circulation | a. a part of the general circulation which is made up of veins and carries digested food and water to the liver |
| _____ | 2. renal vein | b. that part of the general circulation which carries blood from aorta to the kidneys and back to the heart |
| _____ | 3. portal circulation | c. the part of the general circulation which nourishes the heart |
| _____ | 4. renal circulation | d. the great vein of the general circulatory system |
| _____ | 5. hepatic portal vein | e. carries limited amounts of cellular waste back to the heart by way of the inferior vena cava |
| | | f. the vein which goes from the liver to the inferior vena cava heavily supplied with nutrients |

# Unit 16
# PULMONARY CIRCULATION

## OBJECTIVES

- Trace the route of the pulmonary circulation
- Describe the function of the pulmonary circulation
- Compare the general and pulmonary circulatory system

The pulmonary circulation carries blood from the heart to the lungs and back to the heart. The blood carried by the pulmonary artery is deoxygenated blood which is a darker red than when it leaves the lungs. There are also waste products in this blood. One of the waste products carried by the blood from the cells is carbon dioxide. It is exchanged for a new supply of oxygen in the lungs. The now oxygenated blood is bright red.

The pulmonary circulation starts its circuit by leaving the right ventricle of the heart through the pulmonary artery, figure 16-1. The pulmonary artery, carrying deoxygenated blood, bifurcates into a right pulmonary artery going to the right lung, and a left pulmonary artery to the left lung. These two arteries eventually branch out into capillaries inside the lungs. Here, the exchange of carbon dioxide for oxygen takes place. The blood, freshly supplied with oxygen, returns to the heart by way of the right and left pulmonary veins. It enters the left atrium of the heart (the opposite side from which it left). It is now ready to make its circuit throughout the body by way of the general circulation to distribute a fresh supply of oxygen.

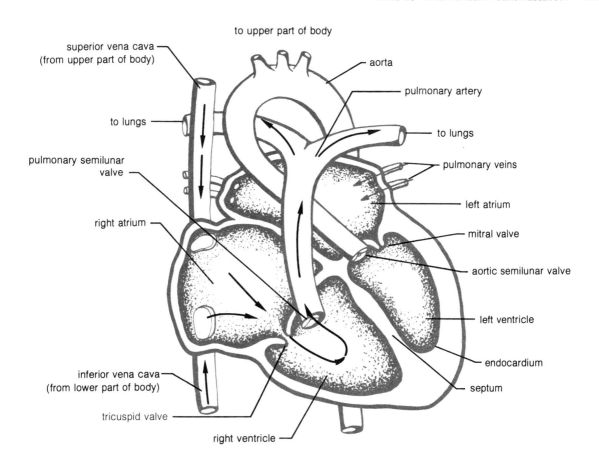

Figure 16-1 Schematic of pulmonary circulation

## Further Study and Discussion

- Research and discuss what is meant by a "blue baby." Why does the baby have a bluish appearance? What can be done for this baby to help it survive?

## Assignment

Briefly answer the following questions.

1. Which two main organs are involved in the pulmonary circulation?

2. Which pulmonary vessel carries deoxygenated blood?

3. Name the major waste material present in deoxygenated blood.

4. What functions do the capillaries perform in the alveoli of the lungs?

5. Name the chamber of the heart from which the blood enters the lungs and the chamber to which the oxygenated blood returns.

# KEY WORDS

arteriole
capillary
diastolic blood
  pressure
endothelial cell
metarteriole
precapillary
  sphincter
pulse
reflux
systolic blood
  pressure
tunica adventitia
  (externa)
tunica intima
tunica media
valves (in
  veins)
vein
venule

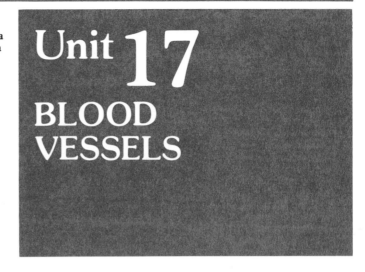

# Unit 17
# BLOOD VESSELS

## OBJECTIVES

- List the five types of blood vessels
- Describe the particular functions of each type of blood vessel
- Identify the principal blood vessels of the body
- Define the Key Words relating to this unit of study

The heart pumps the blood to all parts of the body through a remarkable system of three types of blood vessels: arteries, capillaries, and veins.

## ARTERIES

*Arteries* carry oxygenated blood away from the heart to the capillaries. (There is one exception — the pulmonary arteries — which carry deoxygenated blood from the heart to the lungs). The arteries transport blood under very high pressure; they are thus elastic, muscular and thick-walled. The thickness of the arteries makes them the strongest of the three types of blood vessels.

As seen in color plate 11, the arterial walls are composed of three layers. The outer layer is called the *tunica adventitia,* or *externa.* This layer is composed of loose connective tissue, scattered throughout with bundles of smooth muscle cells. These bundles of smooth muscle cells lend great elasticity to the arteries. This elasticity allows the arteries to withstand sudden large increases in internal pressure, created by the large volume of blood forced into them at each heart contraction. When arteries become hardened, as in *arteriosclerosis,* the *systolic* blood pressure increases greatly.

The *tunica media* is the middle arterial layer. It is composed of muscle cells arranged in a circular pattern. This layer controls the artery's diameter, thus regulating the blood flow through the artery. The tunica media has the most important function, making the arteries very compliant. In this way, the dilation and constriction of the arteries allows for the free flow of blood. This keeps the blood flow steady and even and reduces the heart's work.

111

## 112 SECTION 4 TRANSPORT OF FOOD AND OXYGEN

Table 17-1 Principal Arteries

| NAME | AREA SERVED |
|---|---|
| Common carotid<br>Innominate<br>Right and left coronary<br>Lateral thoracic<br>Pulmonary<br>Aortic arch<br>Right and left subclavian<br>Thoracic aorta | Neck, head and chest |
| Right and left palmar digital<br>Right and left ulnar<br>Right and left radial<br>Right and left brachial | Arms and hands |
| Abdominal aorta<br>Celiac<br>Superior mesenteric<br>Renal | Abdominal region |
| Common iliac<br>Right and left external iliac | Abdomen and legs |
| Right and left deep femoral<br>Right and left femoral<br>Right and left popliteal<br>Right and left anterior tibial<br>Right and left posterior tibial<br>Right and left dorsal pedis | Legs |

An inner layer (*tunica intima*) consists of three smaller layers. The layer lining, facing the lumen of the artery, is made of endothelial cells. They give the arterial lining a smooth surface that does not obstruct blood flow. The next layer consists of delicate connective tissue, found only in arteries with a very large diameter. An elastic layer, composed of a membrane of elastic fibers, is the third layer. This helps to strengthen the arterial walls.

The pulmonary artery and the aorta (the largest artery in the body) lead away from the heart and branch into smaller arteries. These smaller arteries, in turn, branch into *arterioles*. Arterioles give rise to the smallest blood vessels, the *capillaries*, figure 17-1.

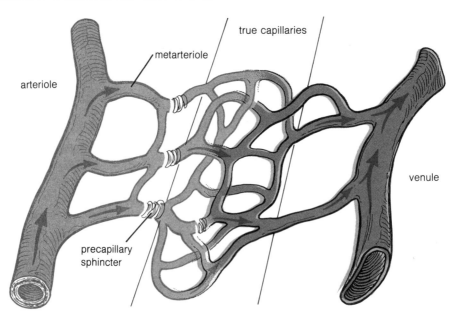

Figure 17-1 Capillary bed connecting an arteriole with a venule

## CAPILLARIES

*Capillaries* are microscopic in size; they are so small they can only be seen through a compound microscope. Capillaries connect the arterioles with tiny veins, called *venules*. Capillaries are branches of the finest arteriole divisions, known as *metarterioles*. The metarterioles have lost most of their connective tissue and muscle layers. Eventually the last traces of these two tissues disappear, and there remains only a simple endothelial cell layer. This endothelial cell layer constitutes the capillaries.

The capillary walls are extremely thin, so as to allow for the selective permeability of various cells and substances. Thus, nutrient molecules and oxygen can pass out of the capillaries and into the surrounding tissues. Consequently, metabolic waste products like carbon dioxide and nitrogenous wastes pass back into the bloodstream for excretion at their proper sites.

Tiny openings in the capillary walls allow white blood cells to leave the bloodstream and enter the tissue spaces to help destroy invading bacteria. In the capillaries, too, some of the plasma diffuses out of the bloodstream and into the tissue spaces. This fluid in the tissue spaces is now called lymph. Lymph is later returned to the bloodstream through the lymphatic vessels.

The internal diameter of a capillary is so small that red blood cells often pass through them in single file. Sometimes the diameter of red blood cells exceeds that of the capillaries; they become compressed and distorted as they flow through the capillary.

Blood flow through the capillaries can be controlled, despite the fact that muscle cells do not line the capillary walls. This is achieved by the action of small muscular bands called *precapillary sphincters.* These bands are found at the region where the capillary branches from an arteriole, or metarteriole.

Although capillaries are ultimately responsible for transporting blood to all tissues, not all capillaries are open simultaneously. This system allows for regulation of blood flow to so-called "active" tissues. In the human brain, for instance, most of the capillaries remain open. However, in a resting muscle, only 1/20 to 1/50 of the capillaries transport blood to the muscle cells. Compare this to an actively contracting muscle where as many as 190 capillaries per square millimeter are open. If the same muscle is not active, there may be as few as five capillaries open per square millimeter.

## VEINS

The *veins* carry deoxygenated blood away from the capillaries to the heart in venules. (A venule is a small vein.) The smallest venules

Table 17-2 Principal Veins

| NAME | AREA SERVED |
|---|---|
| Internal jugular<br>External jugular<br>Right and left innominate<br>Right and left subclavian<br>Pulmonary<br>Superior vena cava<br>Inferior vena cava<br>Right and left axillary | Head, Neck, Chest |
| Right and left cephalic<br>Right and left basilic | Arms |
| Hepatic<br>Portal<br>Splenic<br>Common iliac | Abdominal region |
| Right and left great saphenous<br>Right and left femoral<br>Right and left popliteal<br>Right and left posterior tibial<br>Right and left anterior tibial<br>Right and left dorsal venous arch | Legs |

are hardly larger than a capillary, but they contain a muscular layer which is not present within capillaries.

The vein's structure is comparable to that of the artery; however, veins are considerably less elastic and muscular. Also the walls of the veins are much thinner than those of the arteries since they do not have to withstand such high internal pressures. This is because pressure from the heart's contraction is greatly diminished by the time the blood reaches the veins for its return journey. Thus the thinner walled veins can collapse easily when not filled with blood. Finally, veins have *valves* along their length. These valves allow blood to flow only in one direction, towards the heart. This prevents *reflux* (backflow) of blood toward the capillaries, figure 17-2. Valves are found in abundance in veins, where there is a greater chance for reflux. So, there are many valves in the lower extremities, where blood has to oppose the force of gravity.

Eventually, all the venules converge to make up larger veins, which ultimately form the body's largest veins, the vena cavae. Venous blood from the upper part of the body returns to the right atrium via the superior vena cava; blood from the lower body parts is conducted to the heart via the inferior vena cava.

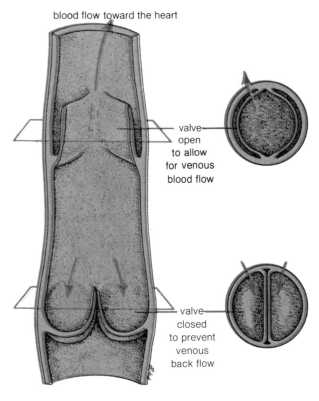

**Figure 17-2 Valves in the vein**

## BLOOD PRESSURE

Initially, when the heart pumps blood into the arteries, the surge of blood filling the vessels creates pressure against their walls. The pressure at the moment of contraction is the *systolic blood pressure,* caused by the rush of blood which follows contraction of the ventricles. The lessened force of the blood (when the ventricles are relaxed) is called *diastolic pressure.* The pressure present in the arteries that are close to the initial surge of blood is greatest and gradually decreases as the blood travels further away from the pumping action.

## PULSE

If you touch certain areas of the body, such as the radial artery at the wrist, you will feel alternating, beating throbs. These throbs represent your body's pulse points. A *pulse* is the alternating expansion and contraction of an artery as blood flows through it.

Try this simple demonstration: place your fingertips (except for the thumb) over an artery which is near the surface of the skin and over a bone. There are six locations where you can conveniently feel your pulse:

- **Brachial artery** – located at the crook of the elbow, along the inner border of the biceps muscle.
- **Common carotid artery** – found in the neck, along the front margin of the sternocleidomastoid muscle, near the lower edge of the thyroid cartilage.
- **Dorsalis pedis artery** – on the anterior surface of the foot, below the ankle joint.
- **Facial artery** – at the lower edge of the mandible, on a line with the corners of the mouth.
- **Radial artery** – at the wrist, on the posterior surface of the radius.
- **Temporal artery** – slightly above the outer edge of the eye.

## Further Study and Discussion

- Investigate and be prepared to discuss the following questions: What is blood pressure? Where is it usually taken? How is it measured? What instruments are used for taking blood pressure?
- Try the following experiment with a classmate and discuss the findings.
    Take the pulse for 30 seconds while the student is seated.
    Take the pulse for 30 seconds while the student is standing.
    Ask the student to jump up and down 25 times, being sure that the knees are well flexed.
    a. Check the pulse immediately after the exercise
    b. Check the pulse one minute after exercise
    c. Check the pulse two minutes after exercise

## Assignment

Match each term in column I with its description in column II.

| Column I | Column II |
|---|---|
| 1. arteries | a. small arteries which lead to capillaries |
| 2. capillaries | b. permit blood to flow in only one direction |
| 3. valves | c. enter the right atrium of the heart |
| 4. veins | d. blood vessels which carry blood back to the heart |
| 5. arterioles | e. connect arterioles with venules |
| | f. large, thick, muscle-walled blood vessels that carry blood away from the heart |

# Unit 18
# THE BLOOD

## KEY WORDS

agglutinin
agglutinogen
agranulocyte (agranular leukocyte)
anticoagulant
antiprothrombin
antithromboplastin
basophil
carboxyhemoglobin
clotting time
coagulation
diapedesis
enucleate
eosinophil
erythrocyte
erythropoeisis
fibrin
fibrinogen
gamma globulin
globin
granulocyte (granular leukocyte)
heme
hemoglobin
leukocyte
lymphocyte
monocyte
neutrophil (polymorphonuclear leukocyte)
oxyhemoglobin
phagocyte
phagocytosis
plasma
prothrombin
Rh factor
serum albumin
serum globulin
thrombin
thrombocyte
thromboplastin
universal donor
universal recipient

## OBJECTIVES

- List the important components of the blood
- Describe the function of each component
- Recognize the significance of the various blood types
- Define the Key Words relating to this unit of study

The average adult has eight to ten pints of blood in his or her body. Loss of more than two pints at any one time leads to a serious condition.

Blood is the transporting fluid of the body. It carries nutrients from the digestive tract to the cells, oxygen from the lungs to the cells, waste products from the cells to the various organs of excretion, and hormones from secreting cells to other parts of the body. It aids in the distribution of heat formed in the more active tissues (such as the skeletal muscles) to all parts of the body. Blood also helps to regulate the acid-base balance, and to protect against infection. Consequently, it is a vital fluid to our life and health.

## BLOOD COMPOSITION

Blood is composed of a liquid portion called *plasma,* which contains various types of dissolved chemicals and several different kinds of blood cells, see color plate 12. Although most of the cells are red blood cells, there are various types of white blood cells as well. The blood also contains large numbers of platelets.

## BLOOD PLASMA

Plasma is a straw-colored, complex liquid, comprising about 55% of the blood volume and containing the following substances in solution:

1. **Water** — Water makes up about 92% of the total volume of plasma. This percentage is maintained by the kidneys, and by water intake and output.

2. **Blood proteins** — There is a protein found in red blood cells known as hemoglobin, which comprises about

two-thirds of the blood proteins. Plasma proteins (discussed next) make up approximately one-third.

3. **Plasma proteins** – These three proteins are the most abundant of those found in plasma: fibrinogen, serum albumin and serum globulin.

   a. *Fibrinogen* is necessary for blood clotting. Without fibrinogen, the slightest cut or wound would bleed profusely. It is synthesized in the liver, and makes up about 4% of the total plasma proteins.

   b. *Serum albumin* is the most abundant of all the plasma proteins. It constitutes some 53% of the total plasma proteins. Another product of the liver, serum albumin helps to maintain the blood's osmotic pressure and volume. It provides the "pulse pressure" needed to hold and pull water from the tissue fluid back into the blood vessels. This is important because metabolic wastes, like carbon dioxide, can be excreted from the cells. Normally, plasma proteins do not pass through the capillary walls, since their molecules are relatively large. Since they are colloidal substances, they can give up, or take up, water-soluble substances, thus regulating the osmotic pressure within the blood vessels.

   c. *Serum globulin* – It makes up about 43% of the total volume of plasma proteins. It is formed not only in the liver, but also in the lymphatic system (discussed in unit 19). *Gamma globulin* has been fractionated (separated) from serum albumin. This portion helps in the synthesis of antibodies, which destroy or render harmless various disease-causing organisms. *Prothrombin* is yet another globulin, formed continually in the liver, which helps blood to coagulate. Vitamin K is necessary in aiding the process of prothrombin synthesis.

4. **Nutrients** – Nutrient molecules are absorbed from the digestive tract. Glucose, fatty acids, cholesterol and amino acids are dissolved in the blood plasma.

5. **Mineral salts** – These are also called *electrolytes,* the most abundant being sodium chloride (NaCl). The others are potassium chloride (KCl), phosphates, sulphates, bicarbonates, etc. These come from foods and chemical processes occurring in the body. They act as chemical buffers in helping to maintain the acid-base balance of the blood.

6. **Hormones, vitamins, and enzymes** – These three substances are found in very small amounts in the blood plasma. They generally help the body to control its chemical reactions.

7. **Metabolic waste products** – All of the body's cells are actively engaged in chemical reactions to maintain homeostasis. As a result of this, waste products are formed and subsequently carried by the plasma to the various excretory organs.

# RED BLOOD CELLS

Red blood cells, or *erythrocytes*, are biconcave, disc-shaped cells. They are caved in on both sides, with a thin center and thicker margins. When viewed from above, they appear to have a doughnut shape, see color plate 12.

## Hemoglobin

Erythrocytes contain a red pigment (coloring agent) called *hemoglobin*, which provides its characteristic color. Hemoglobin is composed of a protein molecule called *globin*, and an iron compound called *heme*. A single red blood cell contains several million molecules of hemoglobin. Hemoglobin is vital to the function of the red blood cell, helping it to transport oxygen to the tissues and carbon dioxide away from the tissues.

## Function

In the capillaries of the lung, erythrocytes pick up oxygen from the inspired air. The oxygen chemically combines with the hemoglobin, forming the compound *oxyhemoglobin*. The oxyhemoglobin-laden erythrocytes circulate to the capillaries of tissues. Here oxygen is released to the tissues, and carbon dioxide is picked up by the hemoglobin, forming *carboxyhemoglobin*. The red blood cells circulate back to the lungs to give up the carbon dioxide and absorb more oxygen. The oxyhemoglobin and carboxyhemoglobin are responsible for the blood's color. Except for pulmonary arteries, blood cells that travel in the arteries carry oxyhemoglobin, which gives blood its bright red color. Except for pulmonary veins, blood cells in the veins contain carboxyhemoglobin, which is responsible for the dark, crimson-blue color characteristic of venous blood.

## Erythropoiesis

Erythropoiesis, or synthesis of red blood cells, occurs in the red bone marrow of essentially all bones, until adolescence. (In the fetus, red blood cells are also produced by the spleen and liver.) As one grows older, the red marrow of the long bones is replaced by fat marrow; erythrocytes are thereafter formed only in the short and flat bones.

Erythrocytes come from primitive cells in the red bone marrow called *hemocytoblasts*. As the hemocytoblast matures into an erythrocyte, it loses its nucleus and cytoplasmic organelles. The hemocytoblast also becomes smaller, gains hemoglobin, develops a biconcave shape, and enters into the bloodstream. To aid in erythropoiesis, vitamin $B_{12}$ and an intrinsic factor are needed. Additional requirements include copper, cobalt, iron, folic acid, proteins and several vitamins.

Since erythrocytes are *enucleated* (contain no nucleus), they only live about 120 days. Destruction occurs as the cells age, rendering them more vulnerable to rupturing. They are broken down by the spleen and liver. Hemoglobin breaks down into globin and heme; the iron content of heme is used to make new red blood cells. The normal count of red blood cells ranges from 5,500,000 to 7,000,000 per cubic millimeter of blood for men; 4,500,000 to 6,000,000 per cubic millimeter for women.

# WHITE BLOOD CELLS

White blood cells are known as *leukocytes*. They are larger than the erythrocytes, ranging from 1 1/4 to 2 times their diameter.

They are granular, translucent, and ameboid in shape. Leukocytes are synthesized in both red bone marrow and in lymphatic tissue.

### Types of Leukocytes

Leukocytes are classified into two major groups of cells: the *granulocytes (granular leukocytes)* and the *agranulocytes (agranular leukocytes)*. This classification is due to their cytoplasmic granules, nuclear structure, and reactions to stains like Wright's stain. Granulocytes are synthesized in red bone marrow from cells called *myeloblasts*. Granulocytes are destroyed as they age and as a result of participating in bacterial destruction. In either case, their life-span is only 24 hours.

There are three types of granulocytes: neutrophils, eosinophils, and basophils. Neutrophils are also called *polymorphonuclear leukocytes*.

*Neutrophils* phagocytize bacteria with lysosomal enzymes in the leukocyte. (*Phagocytosis* is a process that surrounds, engulfs and digests harmful bacteria.) *Eosinophils* phagocytize the remains of antibody-antigen reactions. They also increase in great numbers in allergic conditions, malaria, and in worm infestation. *Basophils* also undergo phagocytosis, and their count increases during chronic inflammation and during the healing from an infection.

Agranulocytes are divided into lymphocytes and monocytes. *Lymphocytes* are further subdivided into *B-lymphocytes,* which are synthesized in the bone marrow, and *T-lymphocytes* from the thymus gland (unit 39). Still others are formed by the lymph nodes and spleen. Their life-span ranges from a few days to several years. They basically help the body by synthesizing and releasing antibody molecules.

*Monocytes* are formed in bone marrow. They assist in phagocytosis, and are able to leave the bloodstream to attach themselves to tissues; here they become *tissue macrophages,* or *histiocytes*. During an inflammation, histiocytes help to wall off and isolate the infected area.

The aforementioned types of leukocytes (basophils, neutrophils, eosinophils, and monocytes) which can undergo phagocytosis are called *phagocytes*. Unlike erythrocytes, they can move through the intercellular spaces of the capillary wall into neighboring tissue. This process is known as *diapedesis*.

A normal leukocyte count averages from 5000 to 9000 cells per cubic millimeter of blood (1 leukocyte for every 100 erythrocytes).

To summarize, leukocytes help protect the body against infection and injury. This is achieved through: (1) phagocytosis and destruction of bacteria, (2) synthesis of antibody molecules, (3) "cleaning up" of cellular remnants at the site of inflammation, and (4) the histiocytic action.

## THROMBOCYTES (BLOOD PLATELETS)

*Thrombocytes* are the smallest of the solid components of blood. They are ovoid-shaped structures, synthesized from the larger *megakaryocytes* in red bone marrow. Thrombocytes are actually fragments of the megakaryocyte cytoplasm, see color plate 12.

The normal blood platelet count ranges from 250,000 to 450,000 per cubic millimeter of blood. Platelets function in the initiation of the blood-clotting process. When a blood vessel is damaged, as in a cut or wound, the vessel's collagen fibers come into contact with the platelets. The platelets are then stimulated to produce sticky projecting structures, allowing them to stick to the collagen fibers. The platelets secrete adenosine diphosphate (ADP), which affects adjacent platelets, making them sticky as well. This reaction occurs countless times, creating a "platelet plug" to stop the bleeding. Subsequently, the blood clotting process follows to "harden" the platelet plug. Old platelets eventually disintegrate in the bone marrow.

## Coagulation

Blood clotting or coagulation is a complicated and essential process which depends in large part on thrombocytes. When a cut or other injury ruptures a blood vessel, clotting must occur to stop the bleeding. On the other hand, unnecessary clotting can clog vessels, cutting off the vital supply of oxygen.

Although the exact details of this process are not clear, there is a general agreement that the following reaction occurs. Whenever a blood vessel or tissue is injured, as in a cut, a substance called *thromboplastin* is produced. An injury to a blood vessel makes the lining rough; as blood platelets flow over the roughened area, they disintegrate, releasing thromboplastin.

Thromboplastin is a complex substance that can only cause coagulation if calcium ions and prothrombin are present. Prothrombin is a plasma protein synthesized in the liver.

The thromboplastin and calcium ions act as enzymes in a reaction that converts prothrombin into *thrombin*. This reaction occurs only in the presence of bleeding, because normally there is no thrombin in the blood plasma.

In the second stage of coagulation, the thrombin just formed acts as an enzyme, changing *fibrinogen* (a plasma protein) into *fibrin*. These gel-like fibrin threads layer themselves over the cut, creating a fine, meshlike network. This fibrin network entraps the red blood cells, platelets, and plasma, creating a blood clot. At first, serum (a pale yellow fluid) oozes out of the cut. As the serum slowly dries, a crust (scab) forms over the fibrin threads, completing the common clotting process.

In order for coagulation to occur successfully, two anticoagulants (substances preventing coagulation) must be neutralized. These are called *antithromboplastin* and *antiprothrombin (heparin)*; they are neutralized by thromboplastin. Thromboplastin is released when platelets disintegrate and tissues are injured.

Prothrombin supply, which is necessary to blood-clotting, is dependent on vitamin K. Vitamin K is a catalyst in the synthesis of

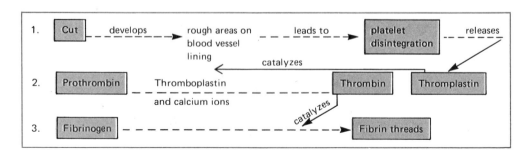

Figure 18-1 Summary of blood clotting reactions

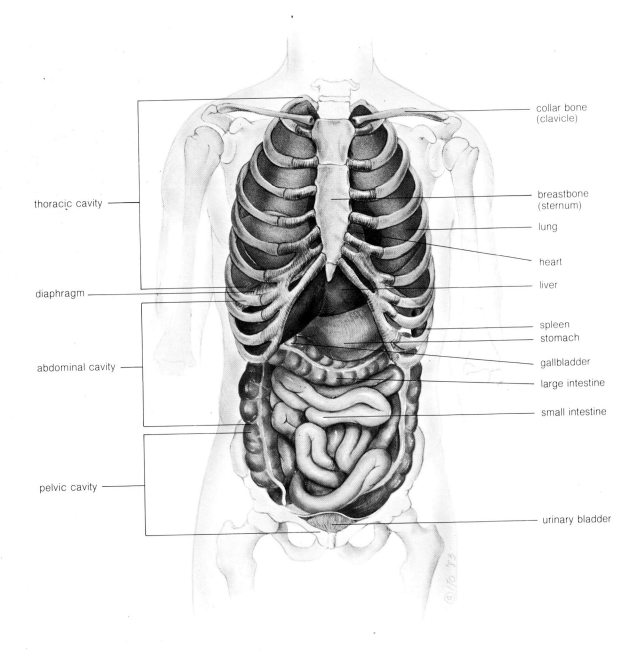

**Plate 1 The ventral cavity**

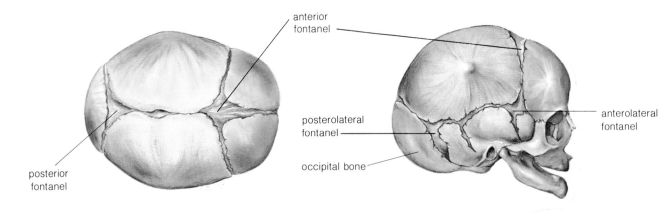

**Plate 2** Top (a) and side (b) views of an infant's skull. The soft cartilage spots on an infant's head are called fontanels.

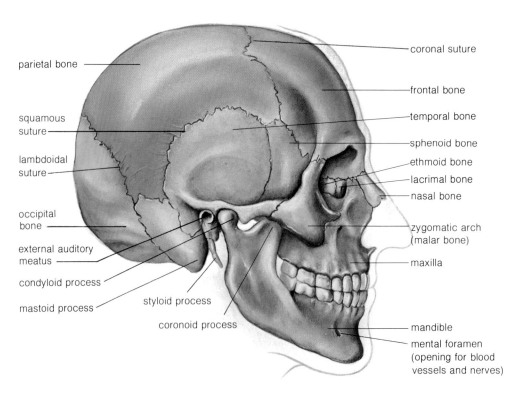

**Plate 3** Side view of the adult skull

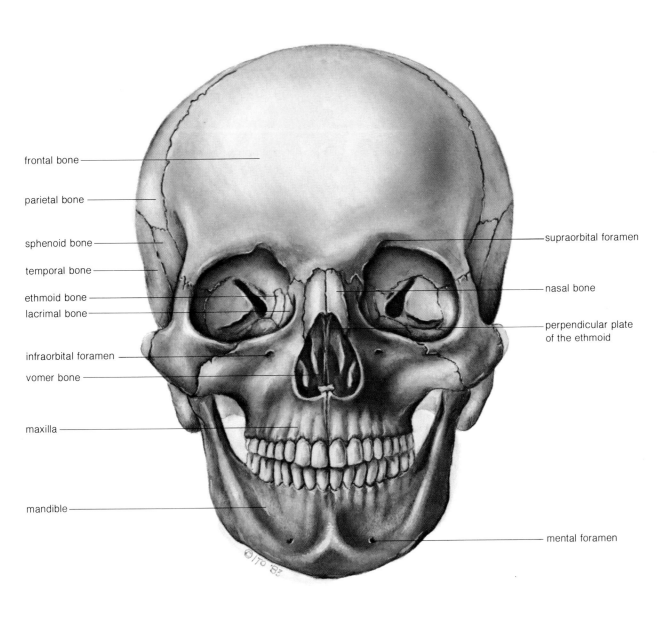

**Plate 4 Front view of the adult skull**

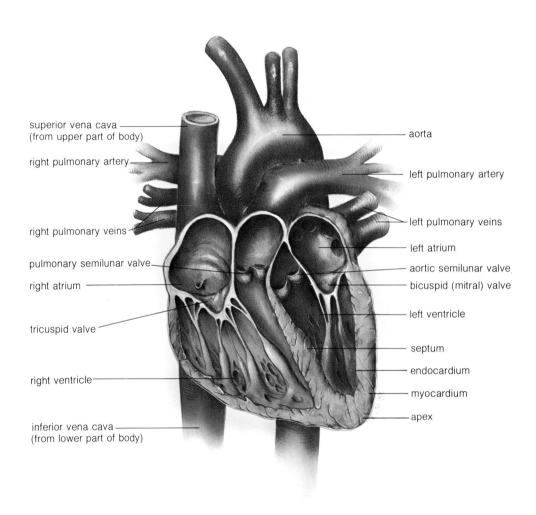

**Plate 5 The heart and its valves**

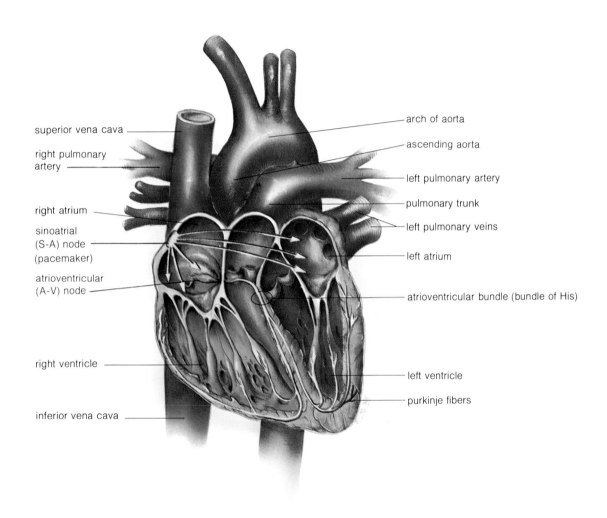

**Plate 6 Conductive pathway of an electrical impulse in a heart contraction**

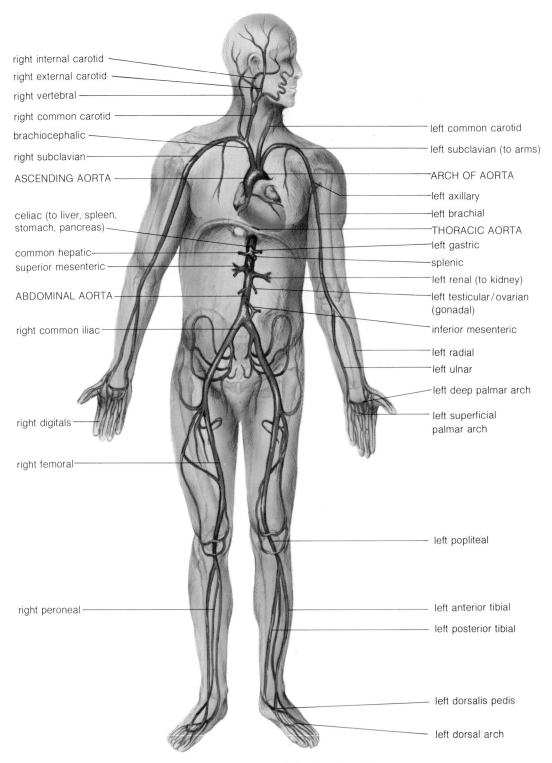

**Plate 7A Arterial distribution**

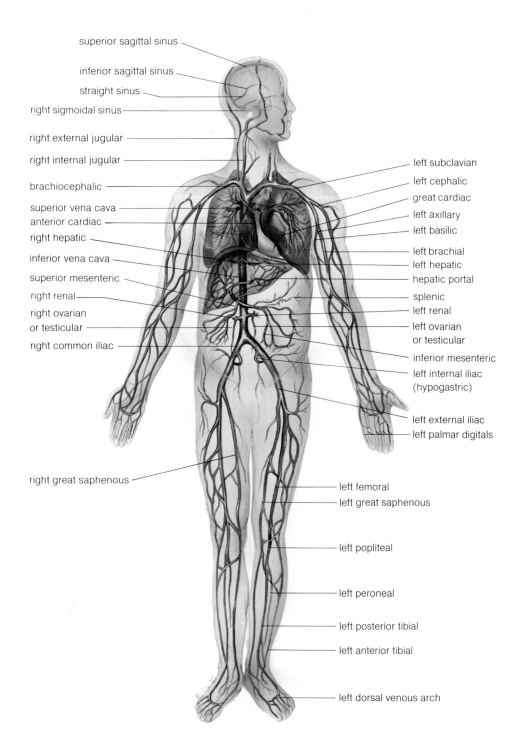

**Plate 7B Venous distribution**

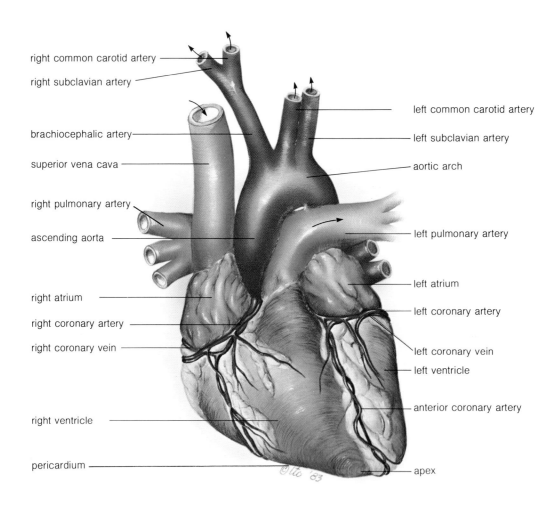

**Plate 8  Front view of the heart**

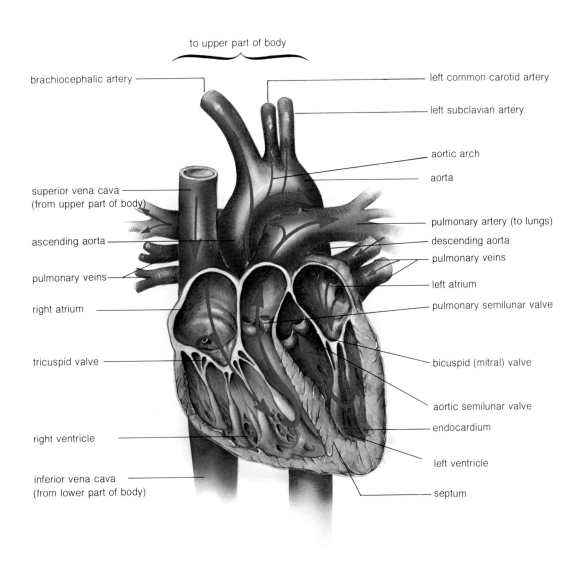

Plate 9  Blood flow into, around, and out of the heart

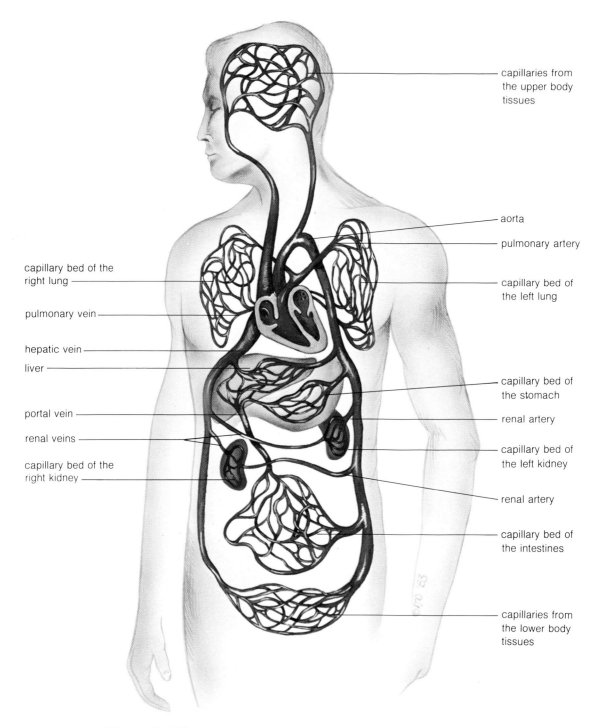

**Plate 10 The systemic, pulmonary, renal, and portal blood circuits**

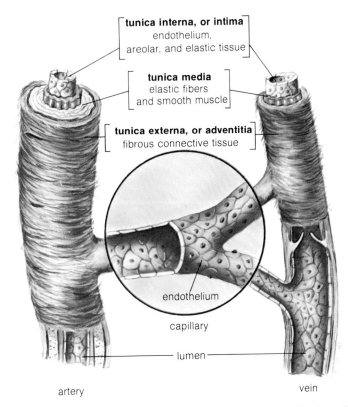

a. Types of blood vessels and their general structure

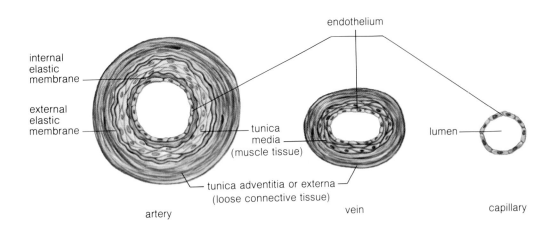

b. Cross section of blood vessels

**Plate 11  Different types of blood vessels and their cross-sectional views**

**a. Red blood cells (erythrocytes)**

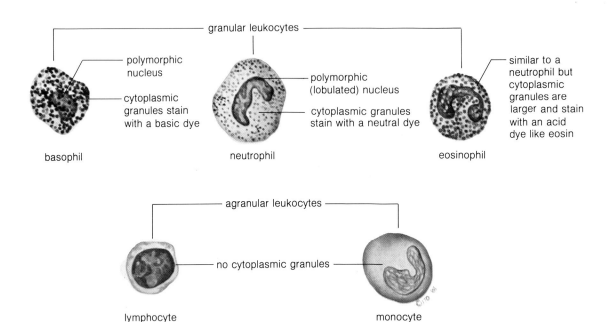

**b. White blood cells (leukocytes)**

**c. Platelets (non-cellular cytoplasmic fragments)**

**Plate 12  Blood cells and platelets**

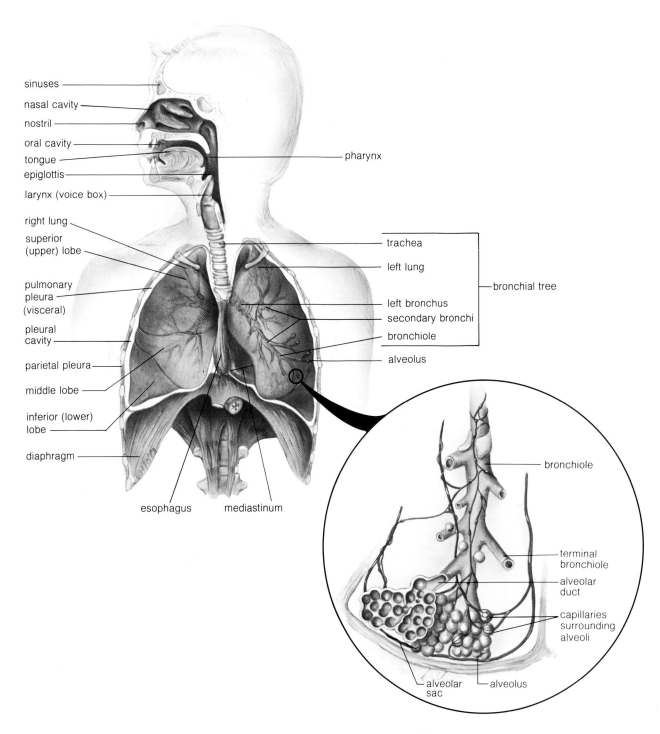

**Plate 13 Respiratory organs and structures**

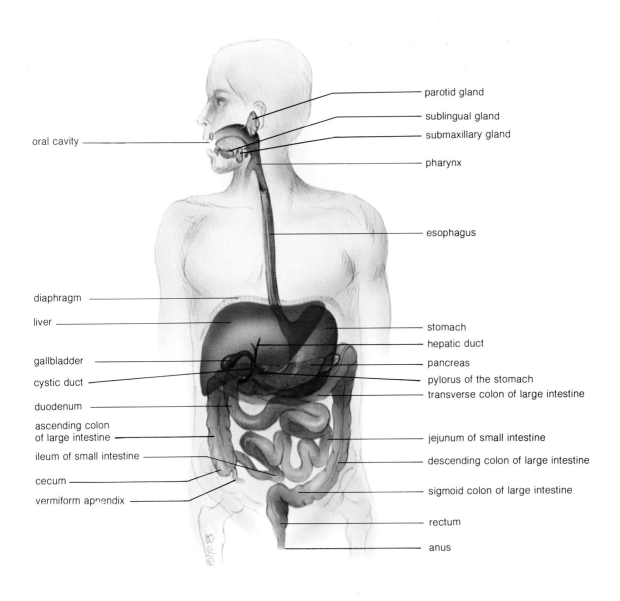

**Plate 14 Alimentary canal and accessory organs**

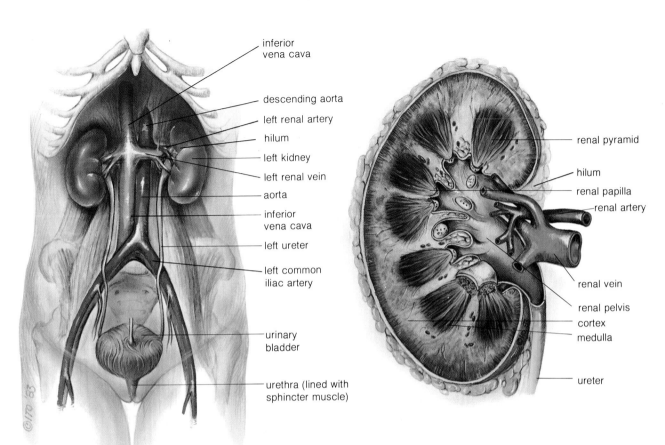

**Plate 15A  The urinary system**

**Plate 15B  Cross section of the kidney**

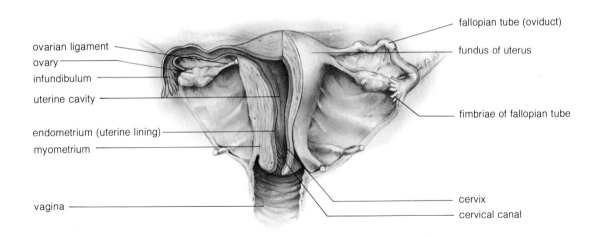

**Plate 16  Uterus, tubes, and ovaries**

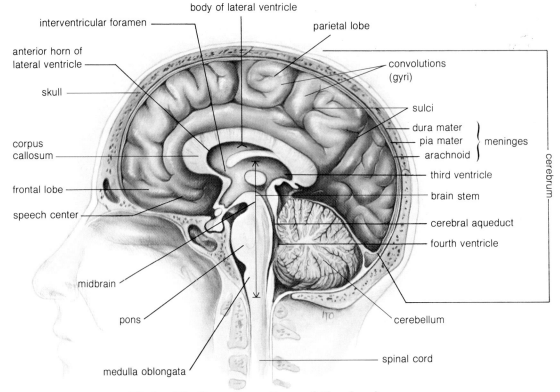

**Plate 17 Cross section of the brain**

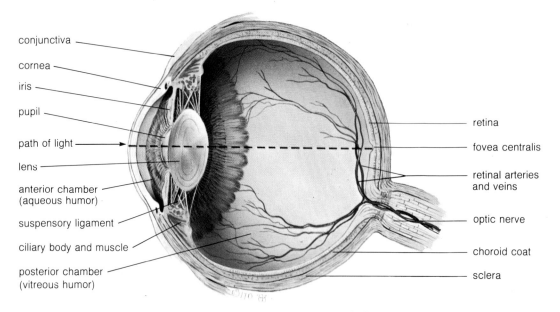

**Plate 18 Internal view of the eye**

prothrombin by liver cells. It is synthesized in the body by a type of bacteria found in the intestines. Some vitamin K may be found in the diet, for example, cabbage, cauliflower, spinach, and soybeans.

**Clotting Time.** The time it takes for blood to clot is known as its *clotting time*. The clotting time for humans is from four to six minutes. This information is quite useful prior to surgery.

## BLOOD TYPES

There are four different groups, or types of blood: A, B, AB and O. Blood type is inherited from one's parents. It is determined by the presence — or absence — of the blood protein called *agglutinogen,* on the surface of the red blood cell. People with type A blood have the A agglutinogen on their red blood cells. Type B blood has the type B agglutinogen; type AB has both A and B agglutinogen; and type O has *neither* of the agglutinogens.

There is a protein present in the plasma, known as *agglutinin*. An individual with type A blood has *b* agglutinin in the blood plasma. Type B blood possesses *a* agglutinin; type O contains *both* a and b agglutinin; and type AB contains *no* agglutinins.

Knowledge of one's correct type is important in cases of blood transfusions and surgery. Agglutinins react with the agglutinogens of the same type, causing the red blood cells to clump together. The clumping of blood clogs up the blood vessels, impeding circulation, thus causing death.

By way of example, if a person with type A blood needs a transfusion, he *must receive only type A blood.* Should he receive type B, the B agglutinogens of the type B blood would clump with the b agglutinins of the person's type A blood. This would prove fatal! However, persons with type A blood can receive both types A and O blood. How is this possible? Because the red blood cells of type O contain no agglutinogens. Therefore they cannot clump with the b agglutinins of type A (or any other). Thus blood type O can be donated to all four blood types, for which reason it is known as the *universal donor.*

Conversely, type AB, having no agglutinins in its plasma, can receive all four blood types. The reason is that, lacking agglutinins, AB cannot agglutinate the red blood cells of any donor. It is thus called the *universal recipient.* Table 18-1 provides a summary of pertinent facts about blood types.

Table 18-1 Blood Types

| BLOOD TYPE | PERCENT OF POPULATION | AGGLUTINOGEN ON RED BLOOD CELLS | AGGLUTININ IN PLASMA | CAN RECEIVE | CAN DONATE TO |
|---|---|---|---|---|---|
| A | 41% | A | b | A or O only | A or AB only |
| B | 12% | B | a | B or O only | B or AB only |
| AB | 3% | A and B | none | A, B, AB, O | AB only |
| O | 44% | None | a and b | O only | A, B, AB, O |

## RH FACTOR

Human red blood cells, in addition to containing agglutinogens A and B, also contain an antigen. This antigen was first discovered in the Rhesus monkey, in 1940, by Karl Landsteiner and Alexander Wierner. We know it as the *Rh factor*. The Rh factor is found within the red blood cell. So people possessing the Rh factor are said to have Rh positive ($Rh^+$) blood. Those without the Rh factor have Rh negative ($Rh^-$) blood.

About 85% of North Americans are Rh positive, 15% Rh negative. Neither Rh negative nor Rh positive blood contains antibodies, or agglutinins in its plasma. However, if an Rh negative individual receives a transfusion of Rh positive blood, he or she will develop antibodies to it. The antibodies take two weeks to develop. Generally there is no problem with the first transfusion. But, if a second transfusion of Rh positive blood is given, the accumulated Rh antibodies will clump with the Rh antigen (agglutinogen) of the blood being received. So, both blood type and Rh factor must be taken into account for safe and successful transfusions.

The same problem arises when an Rh negative mother is pregnant with an Rh positive fetus. The mother's blood can develop anti-Rh agglutinins to the fetus' Rh agglutinogens. The firstborn child will normally suffer no harmful effects. However, subsequent pregnancies will be affected, because the mother's accumulated anti-Rh agglutinins will clump the baby's red blood cells. If the condition is left untreated, the baby will usually be born anemic.

## BLOOD NORMS

Tests have been devised to use physiological blood norms in diagnosing and following the course of certain diseases. Some of these norms are listed in table 18-2.

As stated earlier, prothrombin is a factor found in blood plasma. It is needed for coagulation. The test to determine the prothrombin concentration in the blood plasma is made before and after the administration of vitamin K. If such concentration takes longer to appear than the time shown in table 18-2, liver damage or failure to absorb vitamin K is suspected as a cause.

Table 18-2  Blood Tests

| TEST | NORMAL RANGE |
|---|---|
| Bleeding time | 1 to 3 minutes |
| Coagulation time | 6 to 12 minutes |
| Hemoglobin count | 14 to 16 gms per 100 mL |
| Platelet count | 250,000 to 450,000 per cubic millimeter ($mm^3$) |
| Prothrombin time (quick) | 10 to 15 seconds |
| Sedimentation rate (Westergren) in first hour | Men:    0 to 12 millimeters<br>Women: 0 to 20 millimeters |
| Red blood cell count | Men:    5.5 to 7.0 million/$mm^3$<br>Women: 4.5 to 6.0 million/$mm^3$ |
| White blood cell count | 5000 to 9000/$mm^3$ |

*Sedimentation rate* is the time required for erythrocytes to settle to the bottom of an upright tube at room temperature. It indicates whether disease is present and is very valuable in observing the progression of inflammatory conditions. The Westergren method shows the normal rate for women is slightly higher than for men.

## Further Study and Discussion

- If laboratory facilities are available, practice determining hemoglobin count. Clean the tip of your finger with alcohol (70%). Prick it with a sterile needle. Place a large drop of blood on a piece of filter paper. As soon as the fluid is absorbed, compare it with the color scale which your instructor will have available. What is your hemoglobin? What is the normal hemoglobin count?
- Prepare a slide using a few drops of your own blood. Examine it under the microscope. Note the red and white blood cells.
- Before a transfusion may be given, blood must be typed and cross-matched. Discuss why this is necessary.

## Assignment

A. Briefly answer the following questions.

1. Name the three major types of blood cells.

2. What name is given to the straw-colored liquid portion of the blood?

3. What five proteins are contained in the blood?

4. Which part of the red blood cell is responsible for carrying oxygen?

5. Which body structure is the primary site for red blood cell production?

6. Which type of blood cell protects the body against infection?

7. Which blood cell initiates the blood-clotting process?

8. Name the gel-like substance that forms the tangled threads of the clot.

9. Name six blood tests often relied upon to diagnose and follow the course of blood disorders.

10. Which two organs of the body break down red blood cells?

B. Select the letter which most correctly completes the statement.
   1. Blood of the universal donor is
      a. type B          c. type AB
      b. type A          d. type O
   2. Blood of the universal recipient is
      a. type B          c. type AB
      b. type A          d. type O
   3. Negative Rh blood is found in
      a. 5% of the population     c. 15% of the population
      b. 10% of the population    d. 20% of the population
   4. The blood type found in the largest percent of the population is
      a. type O          c. type AB
      b. type A          d. type B
   5. The prothrombin in the blood-clotting process is dependent upon
      a. vitamin A       c. vitamin P
      b. vitamin K       d. vitamin D

C. Select the most appropriate answer.
   1. One of the following is not a blood cell.
      a. erythrocyte     c. neurocyte
      b. leukocyte       d. phagocyte
   2. Erythrocytes contain all but one of the following elements.
      a. Rh factor       c. hemoglobin
      b. leukocytes      d. globin and heme
   3. What characteristic is not true of normal thrombocytes?
      a. They average 4500 for each cubic millimeter of blood
      b. They are also called platelets
      c. They are plate-shaped cells
      d. They initiate the blood-clotting process

4. The normal leukocyte cell
   a. can only be produced in the lymphatic tissue
   b. goes to the infection site to engulf and destroy microorganisms
   c. is too large to move through the intracellular spaces of the capillary wall
   d. exists in numbers which amount to an average of 12,000 cells per cubic millimeter of blood

5. The blood-clotting process
   a. requires a normal platelet count which is 5000 to 9000 for each cubic millimeter of blood
   b. is delayed by the rupture of platelets which produces thromboplastin
   c. occurs in less time with persons having type O blood
   d. requires vitamin K for the synthesis of prothrombin

# Unit 19
# THE LYMPHATIC SYSTEM

## KEY WORDS

axillary node
inguinal
intercellular (interstitial) fluid
lymph
lymph nodes
lymph vessels
lymphatic system
lymphatics
metastasize
right lymphatic duct
spleen
thoracic duct (left lymphatic duct)
thymus gland

## OBJECTIVES

- Describe the lymphatic system
- Define the components of the lymphatic system
- Outline the function of the lymph nodes
- Define Key Words relating to this unit of study

The *lymphatic system* can be considered a supplement to the circulatory system. It is composed of lymph, lymph nodes, lymph vessels, the spleen, the thymus gland, lymphoid tissue in the intestinal tract, and the tonsils. Unlike the circulatory system, it has no muscular pump or heart.

## LYMPH

Lymph is a straw-colored fluid, similar in composition to blood plasma. Plasma is what diffuses from the capillaries into the tissue spaces. Since lymph bathes the surrounding spaces between tissue cells, it is also referred to as *intercellular* or *interstitial fluid*. Lymph is composed of water, lymphocytes, some granulocytes, oxygen, digested nutrients, hormones, salts, carbon dioxide, and urea. It does not contain any red blood cells or protein molecules too large to diffuse through the capillaries.

Lymph acts as an intermediary between the blood in the capillaries and the tissues. It carries digested food, oxygen, and hormones to the cells. It also carries metabolic waste products (carbon dioxide, urea wastes) away from the cells and back into the capillaries for excretion.

Unlike the circulatory system, the lymphatic system has no pump; other factors operate to push lymph through the lymph vessels. The contractions of the skeletal muscles against the lymph vessels cause the lymph to surge forward into larger vessels. The breathing movements of the body also cause lymph to flow. Valves located along the lymph vessels prevent backward lymph flow.

## LYMPH VESSELS

The lymph vessels accompany and closely parallel the veins. They form an extensive,

**128** SECTION 4 TRANSPORT OF FOOD AND OXYGEN

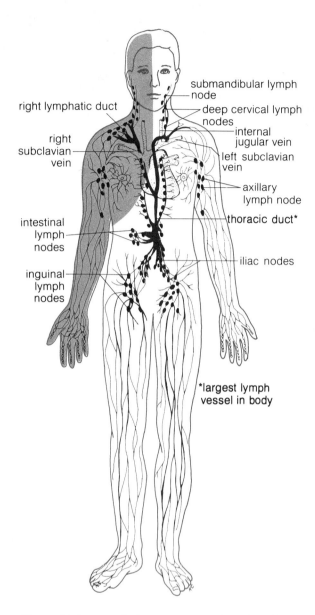

**Figure 19-1** Lymph drainage. Most of the lymph enters the circulation via the thoracic duct but the right lymphatic ducts drain lymph from the right side of the head, right half of the thorax, and the right arm. (From *Human Anatomy and Physiology* By Joan G. Creager. © 1983 by Wadsworth, Inc. Reprinted by permission of Wadsworth Publishing Company, Belmont, California 94002.)

branch-like system throughout the body which may be considered as an auxiliary to the circulatory system.

Lymph vessels are located in almost all the tissues and organs that have blood vessels. They are not found in the cuticle, nails, and hair. Lymphatic capillaries are not in the cartilage, central nervous system, epidermis, eyeball, the inner ear, or the spleen.

The lymph surrounding tissue cells enters small lymph vessels. These, in turn, join to form larger lymph vessels called *lymphatics*. They continue to unite, forming larger and larger lymphatics, until the lymph flows into one of two large, main lymphatics. They are the *thoracic duct* and the *right lymphatic duct*.

The thoracic duct, also called the *left lymphatic duct*, receives lymph from the left side of the chest, head, neck, abdominal area and lower limbs. Lymph in the thoracic duct is carried to the left subclavian vein, and from there to the superior vena cava and the right atrium. In this manner, lymph carrying digested nutrients and other materials can return to the systemic circulation. Lymph from the right arm, right side of the head and upper trunk enters the right lymphatic duct. From there, it enters the right subclavian vein at the right shoulder, then flows into the superior vena cava, figure 19-1.

Unlike the circulatory system, which travels in closed circuits through the blood vessels, lymph travels in only one direction: from the body organs to the heart. It does not flow continually through vessels forming a closed circular route.

## LYMPH NODES

Lymph nodes are tiny, oval-shaped structures ranging from the size of a pinhead to that

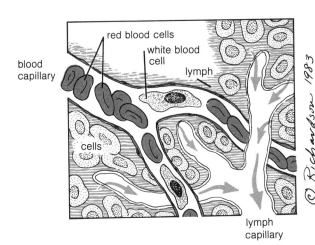

Figure 19-2 Lymph circulation

of an almond, figure 19-2. They are located alone or grouped in various places along the lymph vessels throughout the body. Their function is to make lymphocytes, a type of white blood cell; and antibodies, which serve as a filter for screening out harmful substances (such as bacteria or cancer cells) from the lymph. If the harmful substances occur in such large quantities that they cannot be destroyed by the lymphocytes before the lymph node is injured, the node becomes inflamed. This causes a swelling in the lymph glands, a condition known as *adenitis*.

Knowledge of the location of lymph nodes is important to any health care provider. For example, when giving care to patients with severe infections of the upper leg or thigh, the lymph nodes of the groin (*inguinal*) and the popliteal area are checked for tenderness and swelling.

Another example of care based on knowledge may be applied to patients with breast cancer. In such cases, lymph nodes under the arms (*axillary nodes*) and near the breasts may contain entrapped cancer cells. These cancer cells are filtered out of the lymph that comes from the breast area.

Early detection of unusual lumps in the breast is possible through monthly self-examination. Surgery, chemotherapy, and/or radiation may cure such conditions. But early detection and treatment are *vital*, because if discovered too late, the cancer cells can spread (*metastasize*) to other areas. It is the lymphatic vessels which spread them.

## SPLEEN

The spleen is a sac-like mass of lymphatic tissue. It is located near the upper left area of the abdominal cavity, just beneath the diaphragm. The spleen forms lymphocytes and monocytes. Blood passing through the spleen is filtered, as in any lymph node.

The spleen stores large amounts of red blood cells. During excessive bleeding or vigorous exercise, the spleen contracts, forcing the stored red blood cells into circulation. It also destroys and removes old or fragile red blood cells, and forms erythrocytes in the embryo.

## THYMUS GLAND

The thymus gland is located in the upper anterior part of the thorax, above the heart. Its function is to produce lymphocytes. The thymus is often classified with the lymphatic organs because it is composed largely of lymphatic tissue. It is also considered an endocrine gland because it secretes a hormone which stimulates production of lymphoid cells (this is discussed further in unit 39).

## Further Study and Discussion

- Discuss the diagram in figure 19-2. Explain how materials which the cells need are transported.

## Assignment

Briefly answer the following questions.

1. From what substance is lymph formed?

2. What name is given to the enlarged portion of the lymph vessel?

3. The lymph allows exchange of digested food, oxygen and waste products between two mediums of the body. Name them.

4. What body mechanism forces lymph to move through the lymphatic vessels?

5. Through which vessel does the lymph re-enter the general circulation?

6. Name the two veins which receive lymph drainage at the shoulder.

7. Name two functions of the lymph nodes.

8. Name the inflammatory disorder which produces swelling of the lymph nodes.

9. Identify three body areas where lymph nodes are located.

10. Which of the three major types of blood cells may enter the lymph?

# Unit 20
# REPRESENTATIVE DISORDERS OF THE CIRCULATORY SYSTEM

## KEY WORDS

acute rheumatic heart disease
anemia
aneurysm
angina pectoris
arrhythmia
arteriosclerosis
ascites
atrial fibrillation
cerebral hemorrhage
congenital heart disease
congestive heart failure
Cooley's anemia
coronary occlusion
dyspnea
embolism
endocarditis
familial hemolytic jaundice
gangrene
heart block
heart failure
heart murmur
hemophilia
hemorrhoid
hyperemia
hypertension
hypotension
iron-deficiency anemia
leukemia
microbe
myocardial infarction
myocarditis
pericarditis
pernicious anemia
phlebitis
polycythemia
sickle cell anemia
thrombosis
varicose vein

## OBJECTIVES

- List disorders of the circulatory system
- Describe some disorders of the heart and blood vessels
- Define Key Words related to this unit of study

One of the leading causes of death is cardiovascular disease. Some of the disorders of the heart and blood vessels are defined and briefly discussed in this unit.

## DISORDERS OF THE HEART

*Acute rheumatic heart disease* is an infection of the membrane lining of the heart, usually caused by a streptococcus organism. A streptococcal infection (like a strep throat) may slightly alter heart tissue (e.g., affect heart valves). Subsequent streptococcal infections can cause further heart damage. This is because the antibodies formed against the streptococci also attack the altered heart tissue.

*Arrhythmia* is a term used to describe any change or deviation from the normal orderly rhythm of the heart action.

*Atrial fibrillation* is a condition in which the atria are never completely emptied of blood. Their walls quiver instead of giving the usual contraction of a normal heartbeat. This occurs when irregular and weak nerve impulses arrive at the S-A node, or pacemaker. The pacemaker is stimulated irregularly, and ventricular contraction can occur when it is filled with less than the normal quantity of blood. This causes a highly irregular wrist pulse and an abnormal heart rate.

*Congenital heart disease* is a condition in which the heart did not develop properly during fetal life. Various factors may cause improper heart development during fetal life. For instance, a woman in her second month of pregnancy, who contracts a contagious disease like chickenpox, measles, mumps, rubella (German measles) or smallpox, can pass the pathogenic microorganisms through her placenta into the fetus. The second month is critical, because that is when heart malformations are likely to develop in the fetus. Also, the ingestion of large quantities of aspirin during pregnancy may cause an abnormal heart to develop in the fetus.[1]

*Endocarditis* is an inflammation of the membrane that lines the heart. This causes the formation of rough spots in the endocardium, which may lead to a potentially fatal thrombosis or blood clot.

*Heart block* is the loss of ability of the A-V node to carry nerve impulses. This happens when the A-V node is damaged, possibly due to blockage of a coronary artery leading to the septum. The heart block causes the ventricles to contract at a highly irregular rate, often as slowly as 40 to 50 beats per minute. Heart block is treated by the implantation of a battery-powered pacemaker. This artificial pacemaker stimulates the heart electrically, causing ventricular contractions. Such contractions occur at a sufficiently rapid rate to maintain normal blood circulation.

*Heart failure* is the inability of the heart muscles to beat efficiently due to high blood pressure or other pathological conditions. Different symptoms can arise, depending upon which ventricle fails to beat properly. If the left ventricle fails, *dyspnea* (difficult or rapid breathing) occurs. Engorgement of organs with venous blood, edema and *ascites* (abnormal accumulation of serous fluid in the abdominal cavity) takes place when the right ventricle fails. Other symptoms include lung congestion and coughing.

*Congestive heart failure* is similar to heart failure, but in addition there is edema (swelling) of the lower extremities. Blood backs up into the lung vessels, and fluid extends into the air passages. The patient actually drowns in his or her own fluid.

*Myocarditis* is an inflammation of the heart muscle. An area of the heart muscle may become damaged because of a lack of blood supply, usually due to an occluded artery or a *myocardial infarction*. The symptoms are *hyperemia* (accumulation of blood, leading to distention of blood vessels), cloudy swelling, fatty degeneration, and necrosis. The necrotic (dead) myocardial tissue causes a mild inflammation, resulting in the formation of fibrous tissue.

*Pericarditis* is an inflammation of the membrane covering the heart. The symptoms are *precordial* (chest area overlying the heart) pain and tenderness, cough, dyspnea, rapid pulse and slight fever. It is caused by rheumatic fever, septicemia (poisoning of the blood by the toxic products of invading bacteria), nephritis (kidney inflammation), or the extension of infection from an adjacent area.

*Angina pectoris* is the severe chest pain which arises when the heart does not receive enough oxygen. It is not a disease in itself, but a symptom of an underlying problem with coronary circulation.

The chest pain radiates from the precordial area to the left shoulder, down the arm along the ulnar nerve. Victims often experience an apprehension of impending death. Angina pectoris occurs quite suddenly, brought on by emotional stress or physical exertion. It commonly affects men over 40 years old, but rarely women.

*Murmurs* can indicate some defect in the valves of the heart. They may take the form of "gurgling" and "hissing" sounds as the valves fail to close properly. Cardiac murmurs are designated as bicuspid (mitral), tricuspid, pulmonary or aortic, depending on which valve is damaged. They are also classified according to the period of the heart's cycle

---

1. Sherman and Sherman, *Biology* (New York: Oxford University Press, 1980).

during which they occur. A systolic murmur occurs during systole; the diastolic murmur happens during diastole. A presystolic murmur occurs just before systole.

*Coronary occlusion* is a condition in which the heart does not receive enough blood because a coronary artery is blocked. This heart disease is also known as coronary thrombosis, and may result from a myocardial infarction.

# DISORDERS OF THE BLOOD VESSELS

*Arteriosclerosis* is a thickening of the walls of the arteries. This is caused by fibrous connective tissue, fat deposition, and calcification, causing loss of contractility and elasticity.

*Gangrene* is death of body tissue due to an insufficient blood supply caused by disease or injury.

*Phlebitis* is an inflammation of the lining of a vein, accompanied by clotting of blood in the vein. Symptoms include *edema* (swelling) of the affected area, pain and redness along the length of the vein.

*Varicose veins* are the swollen inelastic veins which result from the slowing up of blood flow back to the heart. Blood backs up in the veins if the muscles do not massage them. The weight of the stagnant blood distends the valves; the continued pooling of blood then causes distention and inelasticity of the vein walls. This condition develops due to hereditary weakness in vein structure. In addition, the human bipedal posture, prolonged periods of standing, and physical exertion can cause valves in the superficial leg veins to enlarge and weaken. Age is yet another factor responsible for varicose veins.

*Hemorrhoids* are varicose veins in the walls of the lower rectum and the tissues around the anus.

*Cerebral hemorrhage* refers to bleeding from blood vessels within the brain. It can be caused by arteriosclerosis, disease, or an injury like a blow to the head.

*Aneurysm* is a sac caused by enlargement of the blood vessel, accompanied by thinning of the vessel wall. The aneurysm also forms a blood-containing tumor, which pulsates with each systole, causing a murmur. The symptoms are pain and pressure. The most common aneurysm occurs in the aorta.

# DISORDERS OF THE BLOOD

*Anemia* is a deficiency in the number of red blood cells. Anemia results from a large or chronic loss of blood (hemorrhage) which decreases the number of erythrocytes. Extreme erythrocyte destruction and malformation of the hemoglobin of red blood cells also causes this condition. Since there is always some hemoglobin deficiency, there is never enough oxygen transported to the cells for cellular oxidation. Consequently, not enough energy is being released. Anemia is characterized by varying degrees of dyspnea, pallor, palpitation, and fatigue.

*Iron-deficiency anemia* is a condition that often exists in children and adolescents. It is caused by a deficiency of adequate amounts of iron in the diet. This leads to insufficient hemoglobin synthesis in the red blood cells. The condition is easily alleviated by ingestion of iron supplements and green, leafy vegetables that contain the mineral, like spinach.

*Pernicious anemia* is a condition affecting erythrocyte development. It leads to both inadequate and abnormally large, misshapen red blood cells. If pernicious anemia is not diagnosed early and continues to progress, death usually occurs. Vitamin $B_{12}$ and the intrinsic factor play a vital role in its prevention and treatment.

*Polycythemia* is a condition in which too many red blood cells are formed. Blood viscosity increases due to friction between the erythrocytes, and also between them and the blood vessel walls. Consequently, there is an increase in blood pressure.

*Embolism* is a condition where an embolus is carried by the bloodstream until it reaches an artery too small for passage. An embolus is a substance foreign to the bloodstream. It may be air, a blood clot, cancer cells, fat, bacterial clumps, a needle, or even a bullet (that was lodged in tissue and breaks free).

*Thrombosis* is a blood clot which forms in a blood vessel – or in the heart – and stays in the same place in which it was formed. The blood clot formed is called a *thrombus*. It is caused by unusually slow blood circulation, or changes in the blood or blood vessel walls.

*Hemophilia* is a hereditary disease in which the blood clots slowly or abnormally. This causes prolonged bleeding with even minor cuts and bumps. Although hemophilia occurs only in males, it is transmitted genetically by females.

*Leukemia* is a condition in which there is a great increase in the number of white blood cells. The overabundant leukocytes replace the erythrocytes, thus interfering with the transport of oxygen to the tissues. They can also hinder the synthesis of new red blood cells from bone marrow. Leukemia is usually fatal. Acute leukemia, which develops quickly and runs its course rapidly, occurs most often in children and young adults.

*Sickle cell anemia* is a chronic blood disease inherited from both parents. The disease causes red blood cells to form in the abnormal crescent shape. These cells carry less oxygen and break easily, causing anemia. The *sickling trait*, a less serious disease, occurs with inheritance from only one parent. Sickle cell anemia occurs almost exclusively among Blacks.

*Cooley's anemia* is a blood disease similar to sickle cell anemia, affecting people of Mediterranean descent.

*Familial hemolytic jaundice* occurs when the spleen functions abnormally, rendering the erythrocytes extremely fragile; they break down quite readily and release their hemoglobin into the blood. This condition can generally be cured by removal of the spleen.

## DISORDERS OF BLOOD PRESSURE

*Hypertension* is high blood pressure in which the systolic reading stays above 140 millimeters of mercury. The average adult reading depends on weight, age, and build.

*Hypotension* is low blood pressure; usually systolic reading is under 100 millimeters of mercury.

---

## *Further Study and Discussion*

- Refer to reading references and other outside sources to find out why phlebitis is dangerous.
- Discuss why patients with damaged hearts must restrict physical activity to limits recommended by their doctor.

## Assignment

Select the correct item which completes each statement.

1. Myocarditis is an inflammation of the
   a. lining of the heart
   b. covering of the heart
   c. arteries of the heart
   d. muscle of the heart

2. Leukemia is a condition in which there is a great increase in
   a. erythrocytes
   b. neurocytes
   c. fibrinogen
   d. leukocytes

3. Hypertension or high blood pressure is a condition in which the systolic reading stays above
   a. 190 mm of mercury
   b. 160 mm of mercury
   c. 140 mm of mercury
   d. 120 mm of mercury

4. Hemophilia is a condition in which the blood clots slowly and is acquired by
   a. environmental contacts and poor nutrition
   b. being inherited by the male but transmitted by the female
   c. close contact with persons who have the disorder
   d. contact with animals in the home and yard

5. A clot which is carried by the blood until it blocks a blood vessel too small for passage is called
   a. an embolism
   b. a thrombus
   c. an aneurysm
   d. a stenosis

# SELF-EVALUATION

## Section 4
## TRANSPORT OF FOOD AND OXYGEN

A. Match each term in column I with its correct description in column II.

| Column I | Column II |
|---|---|
| _____ 1. aorta | a. lower chambers of the heart |
| _____ 2. atria | b. white blood cells which absorb and destroy harmful bacteria |
| _____ 3. cardiac | c. referring to the lungs |
| _____ 4. coronary | d. largest artery in body |
| _____ 5. endocardium | e. liquid part of blood |
| _____ 6. hemoglobin | f. upper chambers of heart |
| _____ 7. lymphatic | g. vessel transporting lymph |
| _____ 8. phagocytes | h. circulation through kidneys |
| _____ 9. pericardium | i. lining of heart |
| _____ 10. portal circulation | j. arteries which nourish heart |
| _____ 11. pulmonary | k. largest vein in body; returns to right atrium |
| _____ 12. plasma | l. oxygen-carrying part of the blood |
| _____ 13. renal | m. pertaining to the heart |
| _____ 14. valves | n. goes to liver from small intestine |
| _____ 15. vena cava (superior and inferior) | o. covering of heart |
| _____ 16. ventricles | p. structures in heart and veins, permit blood flow in one direction only |
| | q. membrane that lines the chest cavity |

B. Select the letter which precedes the correct answer.
   1. An infection of the membrane lining the heart, usually caused by the streptococcus organism is
      a. acute pericarditis
      b. acute myocarditis
      c. acute rheumatic heart disease
      d. acute atrial fibrillation

2. Hypotension is a condition in which the systolic reading usually continues below
   a. 75 millimeters of mercury
   b. 100 millimeters of mercury
   c. 110 millimeters of mercury
   d. 120 millimeters of mercury

3. A condition in which there are too many red blood cells is
   a. pernicious anemia
   b. leukemia
   c. polycythemia
   d. simple or primary anemia

4. A condition in which a blood clot may be carried by the bloodstream until it reaches a blood vessel too small for passage is a (an)
   a. embolism
   b. thrombus
   c. aneurysm
   d. stenosis

5. A condition in which there is an inability to absorb vitamin $B_{12}$ is known as
   a. simple anemia
   b. pernicious anemia
   c. leukemia
   d. polycythemia

C. Match each term in column I with its description in column II.

| Column I | Column II |
| --- | --- |
| 1. adenitis | a. varicose veins in walls of rectum |
| 2. anemia | b. blood clot in artery which leads to the heart |
| 3. angina pectoris | c. inflammation of a vein |
| 4. arteriosclerosis | d. moving blood clot in bloodstream |
| 5. coronary thrombosis | e. inflammation of membrane which lines the heart |
| 6. embolism | f. hardening of the arteries |
| 7. endocarditis | g. severe chest pain |
| 8. hemorrhoids | h. deficiency of red blood cells |
| 9. myocarditis | i. inflammation of heart muscle |
| 10. phlebitis | j. inflammation of lymph glands |
| | k. inflammation of membrane that covers the heart |
| | l. crescent-shaped red blood cells |

# Section 5
# Breathing Processes

# Unit 21
# INTRODUCTION TO THE RESPIRATORY SYSTEM

## KEY WORDS

bicarbonate ions ($HCO_3^-$)
cellular respiration (oxidation)
external respiration
internal respiration

## OBJECTIVES

- Describe the functions of the respiratory system
- List the two stages of respiration
- Describe the mechanics of respiration
- Define the Key Words that relate to this unit of study

The countless millions of cells which make up the human body require a constant supply of energy. This energy is needed to help cells perform their many chemical activities in maintaining the body's homeostasis. For this to occur, energy-rich nutrient (fuel) molecules must be transported to the cells. Oxygen also facilitates the release of energy stored in nutrient molecules. So it must be in constant supply to the body; without oxygen, a human being can live no more than a few minutes at best.

The respiratory system is composed of organs which bring oxygen into the body and remove carbon dioxide. Human respiration is subdivided into three stages: external respiration, internal respiration, and cellular respiration (oxidation).

*External respiration* is also known as breathing, or ventilation. This is the exchange of oxygen and carbon dioxide between the body and the outside environment. The breathing process consists of inhalation and exhalation. As one inhales, the air is warmed, moistened, and filtered on its passage into the air sacs of the lungs (alveoli). The concentration of oxygen in the alveoli is greater than in the bloodstream. Thus oxygen diffuses from the area of greater concentration (the alveoli) to an area of lesser concentration (the bloodstream), then into the red blood cells. Consequently, the concentration of carbon dioxide in the bloodstream becomes greater, and it diffuses into the alveoli. Exhalation expels much of the carbon dioxide in the blood through the alveoli of the lungs. Some water vapor is also given off in the process.

*Internal respiration* includes the exchange of carbon dioxide and oxygen between the cells and the lymph surrounding them, plus the oxidative process of energy in the cells. The differences in concentration of carbon dioxide and oxygen govern the exchange which occurs among the air in the alveoli, the blood, and the tissue cells. After inhalation, the alveoli are rich with oxygen and transfer the oxygen into the blood. The resulting greater concentration of oxygen in the blood moves the oxygen into the tissue cells. Through respiration, the tissue cells use up the oxygen. At the same time, the cells build up a higher carbon dioxide concentration. The concentration increases to a point which exceeds the level in the blood. This causes the carbon dioxide to diffuse out of the cells and into the blood where it is then carried away to be eliminated.

Deoxygenated blood, produced during internal respiration, carries carbon dioxide in the form of bicarbonate ions ($HCO_3^-$). These ions are transported by both blood plasma and red blood cells. Exhalation expels carbon dioxide from the red blood cells and the plasma; it is released from the body in the following manner:

$$H_2CO_3 \longrightarrow H_2O + CO_2$$

(Bicarbonate ions decompose to form water and carbon dioxide)

*Cellular respiration,* or *oxidation,* involves the use of oxygen to release energy stored in nutrient molecules like glucose. This chemical reaction occurs within the cells. Just as wood, when burned (oxidized), gives off energy in the form of heat and light, so does food give off energy when it is burned, or oxidized, in the cells. Much of this energy is released in the form of heat to maintain body temperature. Some of it, however, is used directly by the cells for such work as contraction of muscle cells. It is also used to carry on other vital processes.

As wood burns, the carbon and hydrogen combine with oxygen to form carbon dioxide ($CO_2$) and water vapor ($H_2O$). Similarly, food, when oxidized, gives off waste products including carbon dioxide and water vapor. These are transported from the cells through the circulatory system to the lungs where they are exhaled.

## Further Study and Discussion

- Breathe on a mirror. Note the moisture which appears from the exhaled air. Discuss the fact that carbon dioxide, heat, and water vapor are given off in exhalation.

- Jog in place. Note the effect of body activity on the rate of breathing. How does exercise change breathing? Why?

- Why does a young child breathe more rapidly than an aged person?

## Assignment

A. Briefly answer the following questions.

1. Name and describe the three stages of respiration.

2. Describe the oxidation process. What happens to the energy and waste products released?

B. Match each term in column I with its description in column II.

| | Column I | Column II |
|---|---|---|
| _____ | 1. external respiration | a. exchange of gases between the cells and lymph surrounding them |
| _____ | 2. oxidation | b. carbon dioxide and water vapor |
| _____ | 3. waste products of oxidation | c. air sacs in lungs |
| _____ | 4. alveoli | d. filtering of air |
| _____ | 5. internal respiration | e. combining of food and oxygen in tissues |
| | | f. the inhalation of oxygen and exhalation of carbon dioxide |
| | | g. water vapor only |

# Unit 22
## RESPIRATORY ORGANS AND STRUCTURES

### KEY WORDS

adenoid
alveoli
anterior nares
apex of the lung
base of the lung
bifurcate
bronchi
bronchial tubes
bronchiole
cardiac impression
cilia
fissure of the lung
interpleural space
laryngopharynx
larynx
mediastinum
nasal conchae
nasopharynx
olfactory nerve
oropharynx
pharyngeal arch
pharynx
pleura (parietal and visceral)
pleural cavity
pleural fluid
pleurisy
pneumothorax
pulmones
sinus
surfactant
thoracentesis
tonsil
trachea
turbinate

### OBJECTIVES

- List the organs used in breathing
- Define the structures within the lungs
- Describe functions of parts of the respiratory system
- Define the Key Words that relate to this unit of study

Color plate 13 illustrates how air moves into the lungs through several passageways. The following structures are included: nasal cavity, pharynx, larynx, trachea, bronchi, bronchioles, alveoli, lungs, pleura, and mediastinum.

### The Nasal Cavity

In humans, air enters the respiratory system through two oval openings in the nose. They are called the nostrils, or *anterior nares*. From here, air enters the nasal cavity, which is divided into a right and left chamber, or smaller cavity, by a partition known as the *nasal septum*. Both cavities are lined with mucous membranes.

Protruding into the nasal cavity are three *turbinate,* or *nasal conchae* bones. These three scroll-like bones (superior, middle and inferior concha), divide the large nasal cavity into three narrow passageways. The turbinates increase the surface area of the nasal cavity, causing turbulence in the flowing air. This causes the air to move in various directions before exiting the nasal cavity. As it moves through the nasal cavity, air is being filtered of dust and dirt particles by the mucous membranes lining the conchal and nasal cavity. The air is also moistened by the mucus and warmed by blood vessels which supply the nasal cavity. At the front of the nares are small hairs or *cilia* which entrap and prevent the entry of larger dirt particles. By the time the air reaches the lungs, it has been warmed, moistened, and filtered. Nerve endings providing the sense of smell (*olfactory nerves*) are located in the mucous membrane, in the upper part of the nasal cavity.

The *sinuses,* named frontal, maxillary, sphenoid, and ethmoid, are cavities of the skull in and around the nasal region, figures 22-1 and 22-2. Short ducts connect the

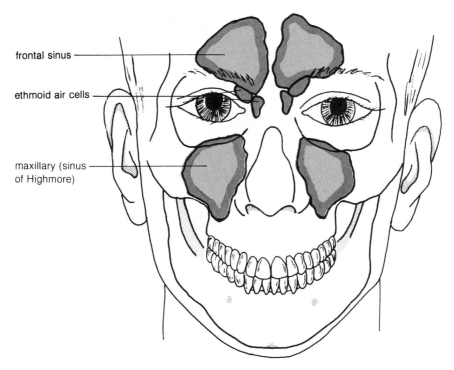

Figure 22-1 Paranasal sinuses, front cross-section view (from P. Anderson, *The Dental Assistant*. Albany: Delmar Publishers Inc.)

sinuses with the nasal cavity. Mucous membrane lines the sinuses and helps to warm and moisten air passing through them. The sinuses also give resonance to the voice. The unpleasant voice sound of a nasal cold results from the blockage of sinuses.

## The Pharynx

After air leaves the nasal cavity it enters the *pharynx,* commonly known as the throat. The pharynx serves as a common passageway for air and food. It is about 5 inches long and can be subdivided into three sections. The uppermost section, just after the nasal cavity, is the *nasopharynx*. The *oropharynx* lies behind the mouth. The lowest portion is known as the *laryngopharynx*. Air travels down the pharynx on its way to the lungs; food travels this route on its way to the stomach. *Adenoids* and *tonsils* are lymphatic tissue, located in the nasopharynx and oropharynx respectively. An upper respiratory infection or cold may cause them to enlarge. Enlargement of the adenoids makes it difficult to breathe. When this happens, breathing is performed through the mouth instead of the nose. The affected individual is said to be "adenoidal."

The left and right eustachian tubes open directly into the nasopharynx, connecting each middle ear with the nasopharynx. Because of this connection, nasopharyngeal inflammation can lead to middle ear infections.

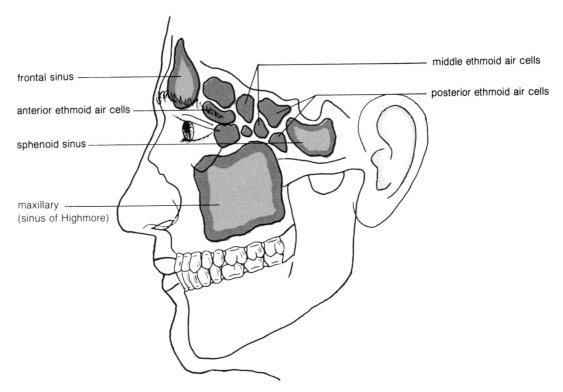

Figure 22-2 Paranasal sinuses, side cross-section view (from P. Anderson. *The Dental Assistant*. Albany: Delmar Publishers Inc.)

## The Larynx

The larynx, or voice box, is a triangular chamber found below the pharynx. The laryngeal walls are composed of nine fibrocartilaginous plates. These plates are derived from embryonic structures called *pharyngeal arches*. Of these nine fibrocartilaginous plates, the largest is the thyroid cartilage, commonly known as the "Adam's apple."

The larynx is lined with a mucous membrane, continuous from the pharyngeal lining above to the tracheal lining below. Within the larynx are the characteristic vocal cords. Muscles attached to the laryngeal cartilages can exert tension upon the vocal cords, lengthening and relaxing them, or making them short and tense. The voice is low-pitched in the former case when the vocal cords are long and relaxed, and high-pitched when they are short and tense.

## The Trachea

The trachea, or windpipe, is a tube-like passageway some 11.2 centimeters (about 4.5 inches) in length. It extends from the larynx, passes in front of the esophagus, and continues to form the two bronchi (one for each lung). The walls of the trachea are composed of alternate bands of membranes, and 15 to 20 C-shaped rings of hyaline cartilage. These C-shaped rings are virtually noncollapsible, keeping the trachea open for the passage of oxygen into the lungs. However, the trachea can be obstructed by large pieces of food, tumorous growths, or the swelling of inflamed lymph nodes in the neck.

The walls of the trachea are lined with both mucous membrane and ciliated epithelium. The function of the mucus is to entrap inhaled dust particles, whereupon the cilia sweep such dust-laden mucus upward to the pharynx. Coughing or regurgitation then dislodges and eliminates the dust-laden mucus from the pharynx.

## The Bronchi and the Bronchioles

The lower end of the trachea *bifurcates* (divides in two) into the *right bronchus* and the *left bronchus.* There is a slight difference between the two bronchi, the right bronchus being somewhat shorter, wider and more vertical in position.

As the bronchi enter the lung, they subdivide into bronchial tubes and smaller bronchioles. The divisions are Y-shaped in form. The two bronchi are similar in structure to the trachea, because their walls are lined with ciliated epithelium and ringed with hyaline cartilage. However, the bronchial tubes and smaller bronchi are ringed with cartilaginous plates instead of incomplete C-shaped rings. The bronchioles lose their cartilaginous plates and fibrous tissue. Their thinner walls are made from smooth muscle and elastic tissue lined with ciliated epithelium. At the end of each bronchiole, there is an alveolar duct which ends in a sac-like cluster called *alveolar sacs (alveoli).*

## The Alveoli

The alveolar sacs consist of many alveoli and are composed of a single layer of epithelial tissue. There are about 300 million alveoli in the adult lung.[1] Each alveolus forming a part of the alveolar sac possesses a globular shape. Their inner surfaces are covered with a lipid material known as *surfactant.* The surfactant helps to stabilize the alveoli, preventing their collapse. Each alveolus is encased by a network of blood capillaries.

It is through the moist walls of both the alveoli and the capillaries that rapid exchange of carbon dioxide and oxygen occurs. In the blood capillaries, carbon dioxide diffuses from the erythrocytes, through the capillary walls, into the alveoli.

Carbon dioxide leaves the alveoli, exhaled through the mouth and nose. The opposite process occurs with oxygen, which diffuses from the alveoli into the capillaries, and from there into the erythrocytes.

If one were to spread out the internal lung surface to form a single sheet of tissue, the entire area would cover an estimated 90 square meters (or 3543.30 square inches).[2] This is more than 100 times the skin surface of the adult human body!

## THE LUNGS

The lungs, or *pulmones,* are fairly large, cone-shaped organs filling up the two lateral chambers of the thoracic cavity, see plate 13. They are separated from each other by the mediastinum and the heart. The upper part of the lung, underneath the collarbone, is the *apex;* the broad lower part is the *base.* Each base is concave, allowing it to fit snugly over the convex part of the diaphragm.

Lung tissue is porous and spongy, due to the tremendous amount of air it contains. If you were to place a specimen of a cow lung

---
1. Morrison, Cornett, Tether, and Gratz, *Human Physiology* (New York: Holt, Rinehart, and Winston, 1977).
2. See note 1 above.

into a tankful of water, for example, it would float quite easily.

The right lung is larger and broader than the left. This is because the heart inclines to the left side. The right lung is also shorter due to the diaphragm's upward displacement on the right in order to accommodate the liver. The right lung is divided by *fissures* (clefts) into three lobes: superior, middle and inferior.

The left lung is smaller, narrower, and longer than its counterpart. It is subdivided into two lobes: superior and inferior. The concave area occupied by the heart, on the left side of the lung, is called the *cardiac impression*.

## The Pleura

The lungs are covered with a thin, moist, slippery membrane made up of tough endothelial cells, or *pleura*. There are two pleural membranes. The one lining the lungs and dipping between the lobes is the *pulmonary*, or *visceral pleura*. Lining the thoracic cavity and the upper surface of the diaphragm is the *parietal pleura*. Consequently, each lung is enclosed in a double-walled sac.

The space between the two pleural membranes is the *pleural cavity*, filled with serous fluid called *pleural fluid*. This fluid is necessary to prevent friction as the two pleural membranes rub against each other during each breath.

Unfortunately, the pleural cavity may, on occasion, fill up with an enormous quantity of serous fluid. This occurs when there is an inflammation of the pleura (called pleurisy). The increased pleural fluid compresses and sometimes even causes parts of the lung to collapse. This obviously makes breathing extremely difficult. To alleviate such pressure, a *thoracentesis* may be performed. This procedure entails the insertion of a hollow, tubelike instrument through the thoracic cavity and into the pleural cavity, so as to drain the excess fluid.

Another disorder which can affect the pleural cavity is *pneumothorax*. This condition occurs if there is a buildup of air within the pleural cavity on one side of the chest. The excess air increases pressure on the lung, causing it to collapse. Breathing is not possible with a collapsed lung, but the unaffected lung can still continue the breathing process.

## THE MEDIASTINUM

The mediastinum, also called the *interpleural space,* is situated between the lungs along the median plane of the thorax. It extends from the sternum to the vertebrae. The mediastinum contains the thoracic viscera: the thymus gland, heart, aorta and its branches, pulmonary arteries and veins, superior and inferior vena cava, esophagus, trachea, thoracic duct, lymph nodes and vessels.

---

### *Further Study and Discussion*

- Using a plastic model or charts, trace the air passage from the nostrils to the air sacs of the lungs.
- If laboratory facilities are available, examine beef lungs and trace the trachea and bronchi into the tissues of the lung. Blow into the trachea with a glass tube and inflate the lungs. Observe the action.

**148** SECTION 5 BREATHING PROCESSES

- Discuss the effect that lung congestion has on breathing (e.g., during a severe chest cold).

## Assignment

A. Explain how the air sacs are particularly adapted to permit a rapid exchange of oxygen and carbon dioxide.

B. Match each term in column I with its correct description in column II.

| Column I | Column II |
|---|---|
| _____ 1. alveoli | a. help give resonance to voice |
| _____ 2. ciliated epithelium | b. air sacs |
| _____ 3. intercostal muscles | c. lines nasal passage |
| _____ 4. larynx | d. voice box |
| _____ 5. lobes | e. divisions of lungs |
| _____ 6. mucous membrane | f. double lining of the thoracic cavity |
| _____ 7. nasal septum | g. chest |
| _____ 8. pleura | h. collects dust particles |
| _____ 9. sinuses | i. made of cartilage |
| _____ 10. thoracic cavity | j. support and aid the breathing process |
| | k. respiratory center |

C. Briefly answer the following questions.

1. What tissue makes up the alveolar sac?

2. What structure forms a tight network around the alveolar sac?

3. What structure, found in the nasal passage and sinuses, will warm, moisten and filter air passing through it?

UNIT 22 RESPIRATORY ORGANS AND STRUCTURES   149

4. Name the four sinuses in the nasal area.

5. Name the two muscles primarily responsible for the breathing movements.

# Unit 23

## MECHANICS OF BREATHING

### KEY WORDS

| | | |
|---|---|---|
| compliance | inspiration | pulmonary |
| expiration | inspiratory reserve | ventilation |
| expiratory reserve | volume | residual volume |
| volume | intercostal muscle | tidal volume |
| fibrotic | medulla oblongata | vagus nerve |
| Hering-Brewer reflex | phrenic nerves | vital capacity |

### OBJECTIVES

- Explain how breathing movements are controlled
- Describe the action of the diaphragm and the ribs in breathing
- Define the various lung capacities
- Define the Key Words related to this unit of study

*Pulmonary ventilation* (breathing) of the lungs is due to changes in pressure which occur within the chest cavity. This variation in pressure is brought about by cellular respiration and mechanical breathing movements.

## THE BREATHING PROCESS

Pulmonary ventilation allows the exchange of oxygen between the alveoli and erythrocyte, and eventually between the erythrocyte and cells.

### Inhalation

There are two groups of intercostal muscles: *external intercostals* and *internal intercostals*. Their muscle fibers cross each other at an angle of 90°. During inhalation, or *inspiration*, the external intercostals lift the ribs upward and outward. This increases the volume of the thoracic cavity. Simultaneously, the sternum rises along with the ribs and the dome-shaped diaphragm contracts and becomes flattened, moving downward. As the diaphragm moves downward, pressure is exerted on the abdominal viscera. This causes the anterior muscles to protrude slightly, increasing the space within the chest cavity in a vertical direction. As a result, there is a decrease in pressure. Since atmospheric pressure is now greater, air rushes in all the way down to the alveoli, resulting in inhalation.

### Exhalation

In exhalation, or *expiration,* just the opposite takes place. Expiration is a passive process; all the contracted intercostal muscles and diaphragm relax. The ribs move down, the diaphragm moves up. In addition, the surface tension of the fluid lining the alveoli

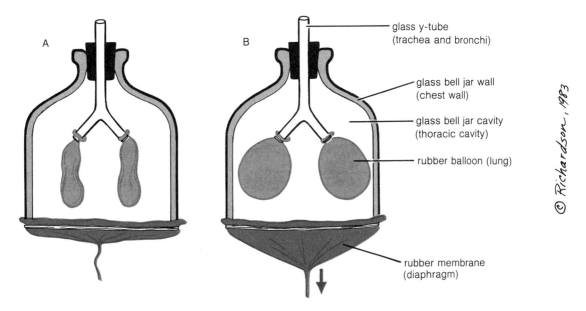

Figure 23-1 Glass bell jar models to demonstrate the process of inhalation and exhalation. Model A represents exhalation — the rubber diaphragm is relaxed, deflating the two rubber balloons (lungs). Model B represents inhalation — the rubber diaphragm is pulled downward, inflating the two rubber balloons. This occurs because the volume of the bell jar cavity has been increased.

reduces the elasticity of the lung tissue and causes the alveoli to collapse. This action, coupled with the relaxation of contracted, respiratory muscles, relaxes the lungs; the space within the thoracic cavity decreases, thus increasing the internal pressure. Increased pressure forces air from the lungs, resulting in exhalation. Figure 23-1 illustrates this concept. Two glass bell jars, equipped with a rubber stopper and balloons, are used to demonstrate the mechanics of breathing.

The lungs are extremely elastic. They are able to change capacity as the size of the thoracic cavity is altered. This ability is known as *compliance*. When lung tissue becomes diseased and fibrotic, the lung's compliance decreases and ventilation increases.

## Pressure Changes During Respiration

During inhalation, as the thoracic cavity expands so do the lungs. Lung expansion lowers the air pressure in the alveoli and air sacs, resulting in a pressure gradient which, in turn, allows air to rush into the alveoli. This helps to equalize the air pressure between the outside and the alveoli, decreasing pressure in the pleural cavity.

In order to understand the concept of gaseous pressure, a brief discussion of atmospheric pressure will prove helpful. The pressure of a gas is measured by the height of a column of mercury which the gas is able to support. Imagine that a glass cylinder, sealed at one end, is filled with mercury; it is then inverted into a shallow receptacle containing mercury. Eventually the mercury in the cylinder drops to where its weight is balanced by the atmospheric pressure exerted on the surface of the mercury in the shallow container. At sea level, the mercury in the cylinder will rise to a height of 760 millimeters, or 30 inches. This figure (760 mm) is known as 1 atmosphere of pressure of air at sea level at zero degree Celsius (0°C or 32°F). At higher elevations, the column of mercury will decrease due to the decrease in atmospheric pressure, and vice-versa. The instrument that measures atmospheric pressure is called a *barometer*.

During inhalation, alveoli pressure decreases in proportion to pressure in the pleural cavity, by about 2–3 millimeters of mercury to 757–758 millimeters of mercury (mm Hg). This slight change in air pressure is enough to push air into the lungs and alveoli. Exhalation causes the alveoli pressure to rise about +3 mm Hg to 763 mm Hg. Consequently air rushes out of the alveoli and lungs.

## Control of Breathing

The rate of breathing is controlled by neural (nervous) and chemical factors. Although both have the same goal — that of respiratory control — they function independently of one another.

**Neural Factors.** The respiratory center is located in the *medulla oblongata* in the brain. It is subdivided into two centers: one to regulate inspiration, the other for expiratory control, see figure 23-2.

The upper part of the medulla contains a grouping of cells that is the seat of the respiratory center. This has been experimentally proven by applying electrical and thermal stimulation to the medulla of research animals. Application of stimuli to this respiratory center causes the rate of respiration to increase or decrease.

Two neuronal pathways are involved in the breathing process. One group of motor nerves, called the *phrenic nerves,* leads to the diaphragm and the intercostal muscles. The other nerve pathway carries sensory impulses from the nose, larynx, lungs, skin, and abdominal organs to the *vagus nerve* in the medulla. As you can see, various stimuli from different parts of the body help to control breathing.

The rhythm of breathing can be changed by stimuli originating within the body's surface membranes. For example, a sudden drenching with cold water can make us gasp, while irritation to the nose or larynx can make us sneeze or cough.

Although the medulla's respiratory center is primarily responsible for respiratory control, it is not the only part of the brain that controls breathing. A lung reflex, called the *Hering-Brewer reflex,*[1] is involved in preventing the overstretching of the lungs. When the lungs are inflated, the nerve endings in the walls are stimulated. A nerve message is sent from the lungs to the medulla by way of the vagus nerve, inhibiting inspiration and stimulating expiration. This mechanism prevents overinflation of the lungs, keeping them from being ripped apart like an over-inflated balloon. Also it prevents the lungs from taking up too much blood, and so depriving the left side of the heart of its blood supply.

**Chemical Factors.** The respiratory center can be inactivated by sleep-inducing drugs, or other central nervous system depressants such as excessive and toxic amounts of barbiturates,

---
1. Karl Hering (1834–1918), German physiologist and Josef Brewer (1842–1925), Austrian psychiatrist.

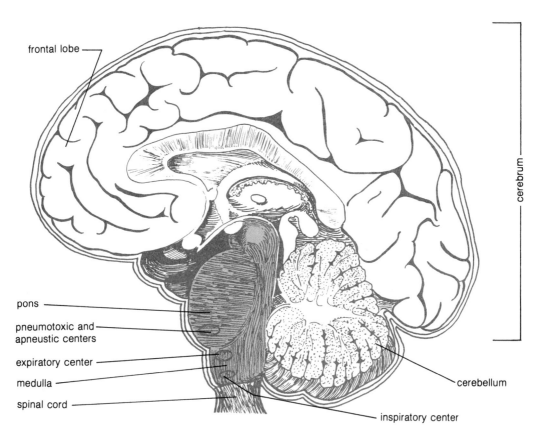

Figure 23-2 Cross section of the brain. Note the expiratory and inspiratory centers.

chloroform, ether, and morphine. These drugs interfere with oxygen utilization by the cells of the respiratory center. They also damage nerve impulse transmissions and decrease the blood flow into the respiratory center. If such damage to the respiratory center is not avoided, death will occur. Currently, there are few chemical substances which can reactivate the respiratory center. Therefore, a "chemical antidote" for overdose of sleep-inducing drugs is not available. In many cases, artificial respiration is used to reactivate the respiratory center.

Chemical control of respiration is dependent upon the level of carbon dioxide in the blood. When blood circulates through an active tissue, it receives carbon dioxide and other metabolic waste products of cellular respiration. As blood circulates through the respiratory center, it becomes sensitive to the increased carbon dioxide in the blood. Consequently, a person performing vigorous exercise or physical labor breathes more deeply and quickly.

## Lung Capacity and Volume

*Vital capacity* refers to the amount of air one can forcibly expire after a maximum inhalation; in other words, it is a measure of the ability to inspire and expire air. Disease processes that weaken the respiratory muscles (such as polio) or decrease the ability of the lungs to expand (such as emphysema) can decrease the vital capacity. The vital capacity of an average person is about 4500 milliliters.

*Tidal volume* (TV) is the amount of air that is inhaled and exhaled during rest. Normal tidal volume is about 500 milliliters.

*Inspiratory reserve volume* (IRV) is the extra volume of air that can be inhaled over and beyond the tidal volume. It is usually about 3000 milliliters.

The *expiratory reserve volume* (ERV) is the amount of air that can be forced by expiration after the end of a normal exhalation. This is about 1100 milliliters.

The *residual volume* (RV) is what remains in the lungs even after a forced expiration; this is about 1200 milliliters.

## Further Study and Discussion

- Demonstrate breathing movements by placing the hands near the base of the ribs on both sides of the chest. Note how the ribs rise and fall during breathing. Then place the hands horizontally across the stomach (at the base of the diaphragm) as inhalation and exhalation take place.
- Discuss the effects of the following on healthful breathing: posture, exercise, deep breathing, mouth breathing, enlarged tonsils and adenoids.
- Using library sources, look up the "iron lung." How does its action compare with natural breathing movements?
- Discuss why a "respirator" machine is properly called a ventilator.

## Assignment

A. Explain the function of:

a. The diaphragm

b. The intercostal muscles

B. Match each term in column I with its function or description in column II.

| Column I | Column II |
| --- | --- |
| _____ 1. respiratory control center | a. opposite of inhalation |
| _____ 2. inspiration and expiration | b. measure of the ability to inspire and expire air |
| _____ 3. vital capacity | c. complemental air |
| _____ 4. exhalation | d. located in the medulla |
| _____ 5. increases respiratory rate | e. occur from 16 to 24 times a minute |
| _____ 6. diaphragm | f. result of increase in carbon dioxide content of the blood |
| _____ 7. intercostal muscles | g. becomes flattened and moves downward during inhalation |
| _____ 8. tidal air | h. air which cannot be forcibly expelled from the lungs |
| _____ 9. residual air | i. corresponds with atmospheric pressure |
| _____ 10. chest cavity space | j. air inhaled and exhaled during rest |
| | k. muscles in between the ribs which contract during inhalation |
| | l. moves upward during inhalation |

# Unit 24
# REPRESENTATIVE DISORDERS OF THE RESPIRATORY SYSTEM

## KEY WORDS

causative agent
complication
crepitation
dysphagia

inflammatory exudate
mucopurulent discharge

predisposing cause
rales

## OBJECTIVES

- Describe some common respiratory diseases caused by a virus or bacteria
- Describe some respiratory disorders unrelated to infectious causes
- Suggest proper health professional care for respiratory ailments
- Define Key Words related to this unit of study

The greatest loss in production hours each year is caused by the common cold. This respiratory infection spreads quickly through the classroom, factory, or business office. It is often the basis for more serious respiratory disease. It lowers body resistance, making it subject to infection. The direct cause of a cold is usually a virus. Indirect causes include: chilling, fatigue, lack of proper food, and not enough sleep. A person who has a cold should stay in bed, drink warm liquids and fruit juice, and eat wholesome, nourishing foods.

## INFECTIOUS CAUSES

The respiratory system is subject to various infections and inflammations caused by bacteria, viruses, and irritants.

*Pharyngitis* is a red, inflamed throat which may be caused by one or several bacteria or viruses. It also occurs as a result of irritants such as too much smoking or speaking. It is characterized by painful swallowing and extreme dryness of the throat.

*Laryngitis* is an inflammation of the larynx, or voice box. It is often secondary to other respiratory infections. It can be recognized by the incidence of hoarseness or loss of voice. The most common form is chronic catarrhal laryngitis. This is characterized by dryness, hoarseness, sore throat, coughing and *dysphagia* (difficulty in swallowing).

*Tonsillitis* is an infection of the tonsils caused by one of several bacteria. Frequent occurrence of this infection, which is accompanied by severe sore throat, difficulty in swallowing, elevation of temperature, chills and aching muscles, may require surgical removal of the tonsils.

*Sinusitis* is an infection of the mucous membrane which lines the sinus cavities. One or several of the cavities may be infected. Pain and nasal discharge are symptoms of this infection which, if severe, may lead to more serious complications. The sinuses affected can be the ethmoid, frontal, sphenoid, and maxillary sinuses.

*Pleurisy* is the result of inflammation of the pleura, the delicate membrane that lines the thorax and folds back over the lung surface. Pleurisy is often a complication of a more severe illness such as pneumonia or tuberculosis. One type is acute pleurisy, which can originate from a lung infection such as pneumonia, or an inflammation in the mediastinum or adjacent structures. *Acute pleurisy* is characterized by fever, painful breathing, and crepitation (production of sounds from the lungs). Another type is *chronic pleurisy*. Here, fibrinous attachments grow between the parietal and visceral pleura surfaces. Chronic pleurisy may be a complication of tuberculosis.

*Bronchitis* is an inflammation of the mucous membrane of the trachea and the bronchial tubes. It may be acute or chronic and often follows infections of the upper respiratory tract. *Acute bronchitis* can be caused by the spreading of an inflammation from the nasopharynx, or by inhalation of irritating vapors. This condition is characterized by a cough, fever, substernal pain and by rales.

*Chronic bronchitis* occurs in middle or old age, characterized by a cough and dry rales. It can result from repeated bouts with acute bronchitis, gout, rheumatism or tuberculosis; it is sometimes a secondary symptom to cardiac or renal disorders.

*Influenza* is an infectious disease characterized by inflammation of the mucous membrane of the respiratory system. The infection is accompanied by fever, a mucopurulent discharge, muscular pain, and extreme exhaustion. An influenza virus is the causative agent, various strains being labelled as strain A, B, etc. Complications such as bronchopneumonia, neuritis, otitis media (middle ear infection), and pleurisy often follow influenza.

*Pneumonia* is an infection of the lung. The alveoli fill up with an inflammatory exudate. Pneumonia is usually caused by bacteria although there may be other causes. Onset is often sudden and is marked by chills and chest pain. When caused by a virus, it is called *viral* or *atypical* pneumonia.

*Tuberculosis* is an infectious disease of the lungs, caused by the tubercle bacillus **Mycobacterium tuberculosis**. The bacterium can affect any organ or tissue of the body. Most related deaths, however, result from tuberculosis of the lungs. Since there are no obvious early signs of this infection, yearly check-ups and chest X rays are highly recommended. If the tuberculosis bacterium is present in the lungs, it will produce lesions containing lymphocytes and epithelioid cells.

*Diphtheria* is a very infectious disease caused by the **Corynebacterium diphtheriae**. The disease affects the upper respiratory tract. It can be recognized by the formation of a false, grayish-white or yellow membrane on the pharynx, larynx, trachea, and tonsils. Locally, diphtheria causes pain, swelling and obstruction. If the diphtheria toxin is circulated through the bloodstream, it leads to cardiac damage, fever, extreme fatigue, occasional paralysis, and all too often – death.

## NONINFECTIOUS CAUSES

Respiratory ailments which are unrelated to infectious causes sometimes develop in the respiratory system.

*Rhinitis* is the inflammation of the nasal mucous membrane causing swelling and increased secretions. Various forms include *acute rhinitis* (the common cold); *allergic rhinitis,* caused by any allergen (more commonly known as hay fever); *atrophic rhinitis,* which is an atrophy of the mucous membrane of the nose, and *chronic rhinitis,* caused by repeated attacks of acute rhinitis and characterized by the presence of dark, putrid crusts.

*Asthma* is a respiratory disorder with symptoms of difficult breathing, wheezing, coughing, the presence of a mucoid sputum, and a feeling of tightness in the chest. Serious episodes of wheezing may occur from emotional stress or from breathing irritants. Emergency care may be needed.

*Atelectasis* is a condition in which the lungs fail to expand normally, due to bronchial occlusion.

*Bronchiectasis* is the dilation of a bronchus caused by an inflammation, accompanied by heavy pus secretion.

*Silicosis* is caused by breathing dust containing silicone dioxide over a long period of time. The lungs become fibrosed which results in a reduced capacity for expansion. Silicosis is also called *chalicosis, lithosis, miner's asthma, miner's disease,* or *grinder's rot.*

*Emphysema* is a noninfectious condition in which the alveoli become overextended, with resulting overinflation of the lungs. Breathing becomes increasingly difficult. Treatment is aimed at relieving the discomfort of the symptoms. At present, there is no cure, only treatment.

*Cancer of the lung* is a malignant tumor which often forms in the bronchial epithelium. The incidence of this condition is high in middle-aged men, especially those who are regular smokers. Early diagnosis is difficult, depending on examinations rather than apparent symptoms. Surgery is often indicated. If done soon enough, it is often successful.

*Cancer of the larynx* is curable if early detection is made of the disorder. It is found most frequently in men over fifty.

Professional health care in many of these respiratory ailments is directed toward maintaining external respiration while making the patient as comfortable as possible. In addition, sufficient rest and proper nourishment are essential.

## Further Study and Discussion

- Explain why sinusitis is more common among people who live at sea level. Explain why the climate in Arizona is usually beneficial to people who have sinusitis.
- Discuss how a minor infection can lower the body's resistance to other infections.
- Discuss how public health measures could help reduce the incidence of emphysema and lung cancer.

## Assignment

1. Describe what should be done for a person who is developing the symptoms of a cold.

2. Name five common respiratory disorders caused by a virus or bacterium.

3. Name three respiratory disorders of a noninfectious nature.

# SELF-EVALUATION

## Section 5 BREATHING PROCESSES

A. Describe the process of external and internal respiration.

(1) External respiration

(2) Internal respiration

B. Explain the function of the hemoglobin in internal respiration.

C. Label the following drawing by inserting the name of the structure to which the line is pointing (e.g., esophagus).

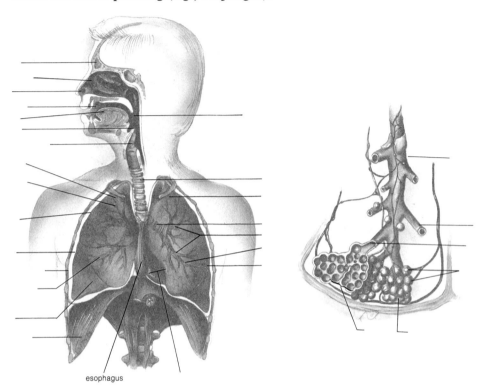

esophagus

D. Complete the following statements.

1. During inhalation, the ribs are _____ by contraction of the rib muscles.
2. The microscopic air sacs in the lungs are called _____.
3. Laryngitis is an inflammation of the _____.
4. When the diaphragm moves _____, inhalation takes place.
5. Energy and waste products are given off when food unites with _____.
6. The principal waste product of exhalation is _____.
7. The respiratory system is dependent upon the _____ system for transporting oxygen to the body cells and carbon dioxide away from the body cells.

8. The lining of the thoracic cavity is called _____.
9. The most common respiratory ailment is _____.
10. The sinuses are lined with _____.

E. Explain how the action of the respiratory center affects the breathing rate.

# Section 6
# Digestion of Food

# Unit 25

# INTRODUCTION TO THE DIGESTIVE SYSTEM

## KEY WORDS

accessory digestive organs
alimentary canal
bicuspids
bolus
buccal cavity
canines
circular muscle
crypts
cusps
deciduous
deglutition
duct of Bartholin
ducts of Rivinus
enzymes
enzymatic hydrolysis
feces
gingivae
hard palate
hydrolytic enzymes
incisors
longitudinal muscle
molars
mucosa
muscle tonus
papillae
• filiform
• fungiform
• circumvallate (vallate)
parotid papilla
peristalsis
salivary glands
• parotid
• sublingual
• submandibular
soft palate
submucosa
tonsils
• lingual
• palatine
uvula
wisdom teeth

## OBJECTIVES

- Describe the general function of the digestive system
- List the structures of the digestive system
- Relate the function of the mouth and teeth to digestion
- Define the Key Words that relate to this unit of study

All food which is eaten must be changed into a soluble, absorbable form within the body before it can be used by the cells. This means that certain physical and chemical changes must take place to *change the insoluble complex food molecules into simpler soluble ones.* These can then be transported by the blood to the cells and be *absorbed through the cell membranes.* The process of changing complex solid foods into simpler soluble forms which can be absorbed by the body cells is called *digestion,* or *enzymatic hydrolysis.* It is accomplished by the action of various digestive juices containing enzymes. *Enzymes* are chemical substances that promote chemical reactions in living things although they themselves are unaffected by the chemical reactions.

Digestion is performed by the digestive system, which includes the alimentary canal and accessory digestive organs. The *alimentary canal* is also known as the digestive tract, gastrointestinal tract (GI tract), or gut. Its organs are found in the head, thorax, abdomen, and pelvis. Most are concentrated within the abdominal cavity. The alimentary canal consists of the mouth (oral cavity), pharynx (throat), esophagus (gullet), stomach, small intestine, large intestine (colon), and the anus, see color plate 14. It is a continuous tube some 30 feet (9 meters) in length, from the mouth to anus. However, during life, the length of the alimentary canal is much shorter (12–15 feet) due to *muscle tonus* (a continual state of partial contraction).

The walls of the alimentary canal are composed of four layers: (1) the innermost lining, called the *mucosa,* is made of epithelial cells, (2) the *submucosa,* consists of connective tissue with fibers, blood vessels, and nerve endings, (3) the third layer is comprised of *circular muscle,* (4) the fourth has *longitudinal muscle.* The mucosa secretes slimy mucus. In some areas, it also produces *hydrolytic (digestive) juices.* This slimy mucus lubricates the alimentary canal, aiding in the passage of food. It also insulates the digestive tract from the effects of powerful hydrolytic enzymes, while protecting the delicate epithelial cells from abrasive substances within the food.

A very important concept must be stressed here regarding the relationship of the alimentary canal to the human body. The body has a "tube-within-a-tube" body plan. The digestive tract can be viewed as the inner tube, the body wall as the outer tube surrounding the digestive tract. In a very real sense, food within the alimentary canal is not yet part of the body and its cells. Thoroughly *digested* food molecules must first pass through the small intestine into the bloodstream, then into the blood capillaries. Here the food molecules diffuse into the lymph, and finally into the body's tissue cells.

The *accessory organs* are the teeth, tongue, salivary glands, pancreas, liver and gallbladder.

## GENERAL OVERVIEW OF DIGESTION

Food enters the gastrointestinal tract via the mouth. In the oral cavity, the food is mechanically digested by the cutting, ripping, and grinding action of the teeth. Chemical digestion of carbohydrates is initiated by the secretion of saliva containing a hydrolytic enzyme. Then, the action of the saliva and rolling motion of the tongue turn the food into a soft, pliable ball called a *bolus.* The bolus slides down to the throat (pharynx) to be swallowed. Next it travels through the esophagus into the stomach. Food is pushed along the esophagus by rhythmic, muscular contractions called *peristalsis.* From the stomach, peristaltic contractions continue to push the food into the small intestine.

Each part of the alimentary canal contributes to the overall digestive process. Protein digestion, for instance, is initiated by the stomach. Then the small intestine starts and finishes fat digestion, as well as completes the digestion of carbohydrates and proteins. Numerous digestive glands are located in the stomach and small intestine, which secrete hydrolytic juices containing powerful enzymes to chemically digest the food. Due to digestion, insoluble food becomes a soluble fluid substance. This substance is then transported across the small intestinal wall into the bloodstream.

Circulated and absorbed through the blood capillaries into the lymph and finally into the body cells, the soluble food molecules are utilized for energy, repair, and production of new cells. The remaining undigested substances (*feces*) pass into the large intestine and leave the alimentary canal via the anus.

## Mouth and Digestion

Food enters the digestive tract through the mouth (oral or *buccal cavity*). The inside of the mouth is covered with a mucous membrane. Its roof consists of a hard and soft palate. The *hard palate* is hard because it is formed from the maxillary and palatine bones, which are covered by mucous membrane. Behind the hard palate is the *soft palate,* made from a movable mucous membrane fold. It encloses blood vessels, muscle fibers, nerves, lymphatic tissue and mucous

**166** SECTION 6 DIGESTION OF FOOD

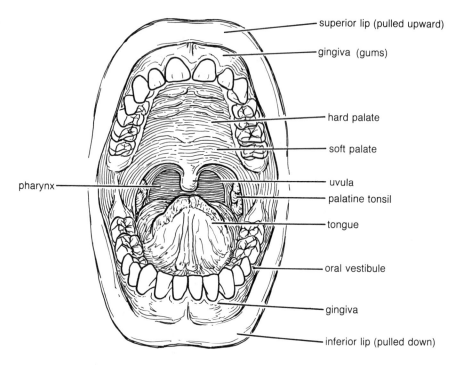

Figure 25-1 The mouth and its structures

glands. The soft palate is an arch-shaped structure, separating the mouth from the nasopharynx. Hanging from the middle of the soft palate is a conical flap of tissue called the *uvula*. This prevents food from entering the nasal cavity when swallowing, figure 25-1.

## Tonsils

Located on either side of the pharyngeal opening are soft masses of lymph tissue known as the *palatine tonsils*. There are also masses of lymph tissue under the tongue called the *lingual tonsils*. Usually, however, the word *tonsils* refers to the palatine tonsils. The tonsils have little openings called *crypts*. Tonsils act like lymph nodes by producing lymphocytes and filtering out microbes that might enter the body and cause infection. Occasionally bacteria may grow inside the crypts, causing an infection which can spread into the lymph and eventually into the blood. This type of infection is known as *tonsillitis*.

## Tongue

The tongue and its muscles are attached to the floor of the mouth, helping in both chewing and *deglutition* (swallowing). The tongue is made from skeletal muscles that lie in many different planes. Because of this, the tongue can be moved in various directions. It is attached to four bones: the hyoid, the mandible, and two temporal bones. On the tongue's epithelial surface are projections called *papillae,* figure 25-2. There are nerve endings located in many of these papillae forming the sense organs of taste, or *taste buds*. These taste buds respond to bitterness, saltiness, sweetness, and sourness in foods,

UNIT 25 INTRODUCTION TO THE DIGESTIVE SYSTEM 167

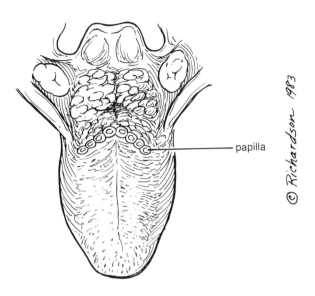

Figure 25-2 There are about 9000 taste buds on the tongue, contained in knoblike elevations called papillae.

figure 25-3(A). They are also sensitive to cold, heat, and pressure. The tongue is made up of three kinds of papillae: *fungiform* on the tip of the tongue, *filiform* near the center and sides, and *vallate,* or circumvallate, near the back, forming a wide "V."

In order for food to be tasted, it must be in solution. The solution passes through the taste bud openings, stimulating the nerve endings in the taste cells. A nerve impulse is created and carried to the brain via nerve fibers at the base of the taste buds, figure 25-3(B).

The sensation of taste is coupled with the sense of smell. When we experience an odor, it stimulates the olfactory nerve endings in the upper part of the nasal cavity. Thus we may confuse the odor of a food with its flavor when it is simultaneously present in the

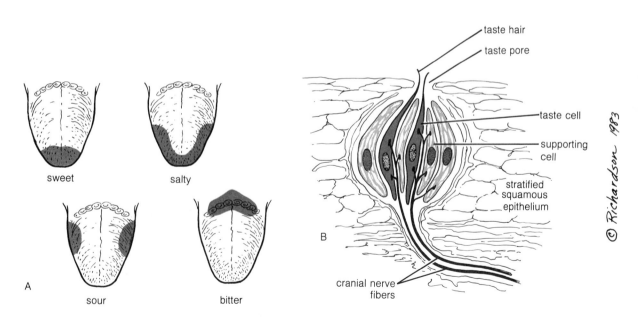

Figure 25-3 Taste buds are sensitive to four basic tastes as shown in (A). Sketch of a taste bud with its nerve fibers is shown in (B).

mouth. A bad cold, with nasal congestion, frequently impedes the ability to taste the flavor of foods. This is because increased mucous secretions cover the olfactory nerve endings.

## Salivary Glands

The salivary glands produce a watery secretion called saliva. Saliva has many functions. It softens and lubricates food, making it easier to chew and swallow. At the same time, the saliva dissolves a portion of the food so it can be tasted. Saliva contains the enzyme *ptyalin* (salivary amylase), which partially digests starches to simpler substances (dextrose). Since food stays in the mouth only for a short time, ptyalin continues its digestion of the dextrose in the stomach.

Food is swallowed in the form of a bolus, which doesn't break up immediately upon reaching the stomach. Therefore the ptyalin's digestive action can continue as long as thirty minutes more. Eventually the stomach's gastric juice, containing hydrochloric acid, will stop the digestive action of the ptyalin. By that point, as much as 75% of the starches in foods like bread, pasta, potatoes, and rice have been broken down.

Saliva also neutralizes mouth acids, washes the teeth, and keeps the mouth cavity flexible and moist. A moist mouth helps in the speech process.

Saliva is a watery mixture of different chemical compounds. It consists of 99% water and contains dissolved traces of salts like calcium, potassium, and sodium, as well as mucin and the enzyme ptyalin.

Saliva is secreted into the oral cavity by three pairs of salivary glands: the parotid, the submandibular, and the sublingual, figure 25-4. The *parotid salivary glands* are found on both sides of the face, in front and below the ear. They are the largest salivary glands, the ones that become inflamed during an attack of mumps. Chewing, at such times, is painful, because the motion squeezes these tender, inflamed glands. A parotid duct, also called Stenson's duct,[1] carries its secretion (almost entirely ptyalin) into the mouth. It opens upon the inner surface of the cheeks, opposite the second molar of the upper jaw. This area is marked by a flap of tissue called the parotid papilla.

Below the parotid salivary glands and near the angle of the lower jaw is the *submandibular gland*. This gland is about the size of a walnut, and its secretions contain both mucin and ptyalin. The secretions enter the buccal cavity via the submandibular duct, or Wharton's duct,[2] at the anterior base of the tongue.

The final pair of salivary glands are the *sublingual glands*, the smallest of the three. They are found under the sides of the tongue. Their secretion consists mainly of mucus and contains no ptyalin. There are some 8-20 small ducts (*ducts of Rivinus*), and a larger one (*duct of Bartholin*), which either join the submandibular duct or open directly onto the mouth floor. An infected sublingual gland causes swelling of the floor of the mouth, resulting in pain during any motion of the tongue, such as in chewing and talking.

## Gingivae and Teeth

The *gingivae*, or gums, support and protect the teeth. They are made up of fleshy tissue covered with mucous membrane. This membrane surrounds the narrow portions of

---

1. Nicholas Stenson (1638-1686), Danish anatomist
2. Thomas Wharton (1610-1673), English anatomist

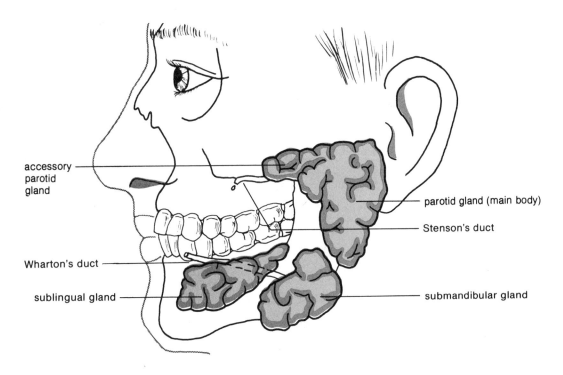

Figure 25-4 Salivary glands (from P. Anderson, *The Dental Assistant.* **Albany:** Delmar Publishers Inc.)

the teeth (also called cervix or neck), and covers the structures in the upper and lower jaws.

Food ingested by the mouth must be thoroughly chewed, or *masticated,* by the teeth. Teeth help break food down into very small morsels, increasing the food's surface area. This activity enables the digestive enzymes to digest the food more efficiently and quickly than if it were swallowed without being chewed. During normal growth and development, the human mouth develops two sets of teeth: (1) the deciduous or milk teeth, which are later replaced by (2) the permanent teeth.

*Deciduous teeth* start to erupt at about 6 months and continue until around two years of age. In total, 20 deciduous teeth are cut during the first two years — 10 in the upper and 10 in the lower jaw. They are: 4 incisors, 2 canines, and 4 molars. This relationship is expressed in the *dentition formula* as shown in figure 25-5(A). The *incisors* have sharp edges for biting, the *canines* are pointed for tearing, and the *molars* have ridges, or *cusps,* designed for crushing and grinding. There are no *premolars* among the deciduous teeth. Deciduous teeth last only up to the age of six.

*Permanent teeth* begin developing at this point, pushing out their deciduous predecessors. The first molars lead the way between the fifth and seventh years. The last to emerge are the third molars, or *"wisdom teeth,"* which may appear anywhere from 17 to 25 years of age. In total, the adult mouth develops 32 teeth, 16 in each jaw, figure 25-5(B).

## 170 SECTION 6 DIGESTION OF FOOD

A. The dentition formula for deciduous teeth is:

|           | Molars | Canine | Incisors | Canine | Molars |
|-----------|--------|--------|----------|--------|--------|
| Upper jaw | 2      | 1      | 4        | 1      | 2      |
| Lower jaw | 2      | 1      | 4        | 1      | 2      |

B. The dentition formula for permanent teeth is:

|           | Molars | Premolars | Canine | Incisors | Canine | Premolars | Molars |
|-----------|--------|-----------|--------|----------|--------|-----------|--------|
| Upper jaw | 3      | 2         | 1      | 4        | 1      | 2         | 3      |
| Lower jaw | 3      | 2         | 1      | 4        | 1      | 2         | 3      |

Figure 25-5 Dentition formulas for deciduous and permanent teeth

Based on the dentition formula, the adult mouth has 8 premolars, or *bicuspids:* 4 in the upper and 4 in the lower jaw. Bicuspids are broad, with two cusps on each crown, and have only one root. Their design is ideal for grinding food. Figure 25-6 shows the arrangement of the deciduous and permanent teeth, and the years during which they normally erupt.

## Further Study and Discussion

- Discuss what measures are necessary for the proper care of teeth.
- Make a list of foods which are rich in calcium and phosphorus.
- Discuss other nutrients important in maintaining the health of the mouth.

## Assignment

A. Briefly answer these questions.

1. What is the purpose of the salivary glands?

2. Name four kinds of permanent teeth, and give the uses of each.

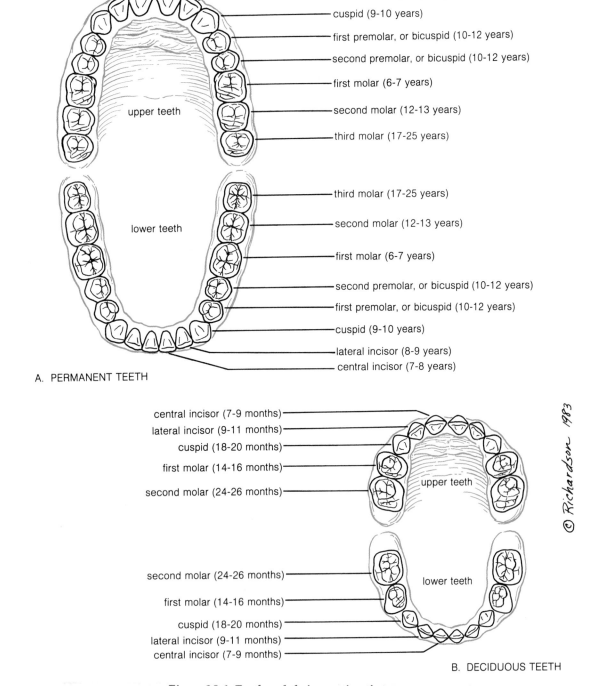

Figure 25-6 Teeth and their eruption times

B. Match each of the terms in column I with its correct statement in column II.

| Column I | Column II |
| --- | --- |
| _____ 1. papillae | a. substances that promote chemical reactions in living things |
| _____ 2. calcium and phosphorus | b. bleeding gums |
| _____ 3. digestion | c. a small soft structure suspended from the soft palate |
| _____ 4. the teeth | d. gums which protect the teeth |
| _____ 5. enzymes | e. tract consisting of the mouth, stomach and intestines |
| _____ 6. gingivae | f. minerals contained in teeth |
| _____ 7. accessory organs and structures of digestion | g. teeth, tongue, salivary glands, pancreas, liver, gallbladder and appendix |
| _____ 8. ptyalin | h. hardest structure in the body |
| _____ 9. uvula | i. projections on the surface of the tongue containing the taste buds |
| _____ 10. alimentary canal | j. the process of changing complex solid foods into soluble forms to be absorbed by cells |
| | k. number of teeth in the adult set |
| | l. the enzyme manufactured by the salivary glands |

C. Label the teeth indicated on the diagrams. (The teeth on the left are the deciduous ones; those on the right are permanent teeth.)

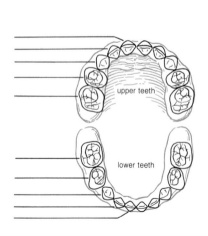
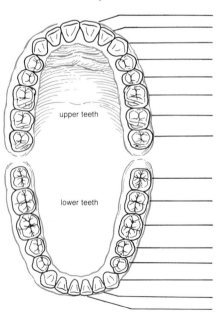

# KEY WORDS

cardiac sphincter
casein
chief cells
deglutition
duodenum
epiglottis
fundus
gastric glands
greater curvature (body of stomach)
greater omentum
intrinsic factor
laryngeal pharynx
mucin
mucous neck cells
mucus
nasopharynx
oropharynx
parietal cells
pepsin
pepsinogen
peptone
peristalsis
pharynx
protease
proteose
pyloric sphincter
pyloric stenosis
pylorospasm
pylorus
rennin
stomach or gastric serosa

# Unit 26
# DIGESTION IN THE STOMACH

## OBJECTIVES

- Describe functions of the pharynx, esophagus and stomach
- Explain the action of gastric juice
- Describe the work of various enzymes
- Define the Key Words that relate to this unit of study

After having been chewed and moistened with saliva in the mouth, food is swallowed by the action of the *pharynx*. The walls of the pharynx consist of skeletal muscle layers lined with mucous membrane. The pharynx is subdivided into three parts: the upper, or *nasopharynx;* the middle, or *oropharynx;* and the *laryngeal pharynx* which reaches from the hyoid bone to the esophagus. The pharyngeal passageway communicates with the ears, nose, mouth, and larynx. In addition, the mucous membrane lining the pharynx is continuous with the mucous membranes of the ears, nasal cavities, mouth, and larynx. The pharyngeal walls are well-lined with mucous glands.

## DEGLUTITION

Swallowing, or *deglutition,* is a complex process involving the constrictor muscles of the pharynx. It begins as a voluntary process, changing to an involuntary process as the food enters the esophagus. When we swallow, the tip of the tongue arches slightly and moves backward and upward. This action forces the food against the hard palate; simultaneously, the soft palate and the uvula shut off the opening to the nasopharynx. Food is thus prevented from entering the nasopharynx.

In swallowing, the constrictor muscles of the pharynx contract, pushing food into the upper part of the esophagus. At the same time, other pharyngeal muscles raise the pharynx, causing the *epiglottis* to cover the larynx (windpipe) to prevent food from entering it. As a further precaution, the nerve impulses causing breathing stop so that breathing and swallowing cannot occur simultaneously. For these reasons, food can only travel in one direction — towards and down the esophagus.

The act of swallowing is voluntary. But, as a bolus of food passes over the posterior part of the tongue and stimulates receptors in the walls of the pharynx, swallowing becomes an involuntary reflex action. With the contraction of the pharyngeal muscles, followed by the contraction of the muscles lining the esophagus, food passes down into the stomach.

## THE ESOPHAGUS

When food is swallowed it enters the upper portion of the esophagus. The esophagus is a muscular tube about 25 centimeters (10 inches) long. It begins at the lower end of the pharynx, behind the trachea. It continues downward through the mediastinum, in front of the vertebral column, and passes through the diaphragm. From there the esophagus enters the upper part, or cardiac portion, of the stomach. This point can be located at the end of the sternum, near the level of the xiphoid process.

The esophageal walls consist of four coats or layers. From the innermost layer to the outermost, they are:

1. the *internal mucous layer*
2. the *submucous* or *areolar layer*
3. a *muscular layer*
4. an *external serous layer* of tough, fibrous connective tissue.

The muscular layer is composed of external longitudinal muscle and internal circular muscle. The muscles in the upper third of the esophagus are striated, and the lower portion consists exclusively of smooth muscle.

After food is swallowed, a series of wave-like involuntary muscular contractions, called *peristalsis,* moves a bolus of food down the esophagus to the stomach. From the time it is swallowed, it takes food five to six seconds to travel the length of the esophagus. When the peristaltic wave reaches the stomach, the *cardiac sphincter* muscle near the entrance to the stomach (cardiac portion) relaxes. This allows food to pass from the esophagus into the stomach. Once the food is in the stomach the cardiac sphincter contracts, preventing reflux, or backflow of food into the esophagus.

## THE STOMACH

The stomach is found in the upper part of the abdominal cavity, just to the left of and below the diaphragm. It is an elastic bag generally shaped like the letter *J*. The shape and position are determined by several factors. These include the amount of food contained within the stomach, the stage of digestion, the position of a person's body, and the pressure exerted upon the stomach from the intestines below.

The stomach is divided into three portions: the upper part or *fundus;* the middle section, called the *body* or *greater curvature;* and the lower portion called the *pylorus.* At the opening into the stomach is found a circular layer of muscle (*cardiac sphincter*) which controls passage of food into the stomach. It is called the cardiac sphincter because of its proximity to the heart. Toward the other end of the stomach lies the *pyloric sphincter* valve which regulates entrance of food into the *duodenum* (the first part of the small intestine). Sometimes the pyloric sphincter valve fails to relax in infants. In such cases, food remaining in the stomach does not get completely digested and eventually is vomited. This condition is called *pylorospasm.* Another abnormal condition is *pyloric stenosis,* a narrowing of the pyloric sphincter which may occur at any age.

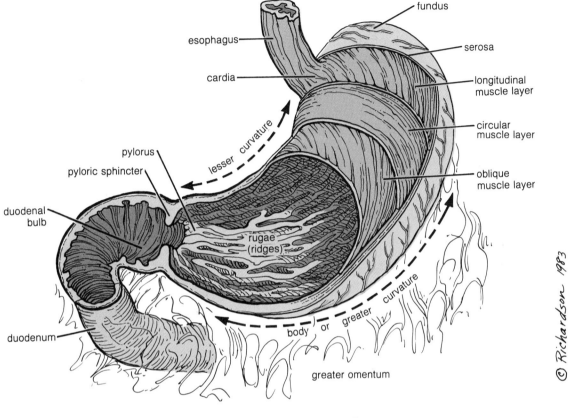

Figure 26-1 Parts of the stomach

The stomach wall consists of four layers: mucous, submucous, muscular, and serous layers.

1. The *mucous coat* is the innermost layer. It is an extremely thick layer made up of small *gastric glands* embedded in areolar connective tissue. When the stomach is not distended with food, the gastric mucosa is thrown into folds called *rugae*, figure 26-1.

2. The *submucosa coat* is made of loose areolar connective tissue. It connects the mucous coat to the muscular coat.

3. The *muscular coat* consists of three layers of smooth muscle: the outer, longitudinal layer; a middle, circular layer; and an inner, oblique layer, figure 26-1. These muscles help the stomach to undergo peristalsis which helps in digestion and in pushing food into the small intestine. The muscular coat is closely connected to the serosa.

4. The serosa is the thick outer layer covering the stomach. It is continuous with the *peritoneum* (the membrane lining the entire inner abdominal cavity). The serosa and peritoneum meet

at certain points, surrounding the organs around the stomach and holding them in a kind of sling. From the left, or greater curvature of the stomach, the peritoneum extends downward, forming the apron-like *greater omentum* which hangs in front of the intestines. The greater omentum contains large amounts of fat.

## The Gastric Glands

The gastric mucosa is estimated to have approximately 35 million gastric glands lining its surface. These glands secrete gastric juice, which is used in digestion.

The gastric glands are comprised of three types of cells: *chief cells, mucous neck cells,* and *parietal cells*, figure 26-2. The chief cells secrete *pepsinogen,* an inactive form of the enzyme *pepsin.* Mucus cells secrete *mucus* and *mucin,* a protein-like substance. The parietal cells secrete hydrochloric acid and the *intrinsic factor.* This is a hormone that helps the body to absorb vitamin $B_{12}$. If insufficient intrinsic factor is secreted, a condition known as pernicious anemia results (see unit 20).

## Gastric Juice

The millions of gastric glands which line the stomach secrete a gastric juice. This gastric juice consists of water, hydrochloric acid, mucus, and the *protease* pepsin (an enzyme that specifically digests proteins).

Hydrochloric acid gives gastric juice a very acidic pH (about 1.5). The functions of hydrochloric acid are:

- It helps to convert the inactive enzyme, pepsinogen, into its active form, pepsin.
- It provides the necessary acidic environment so that the pepsin can begin to digest protein.

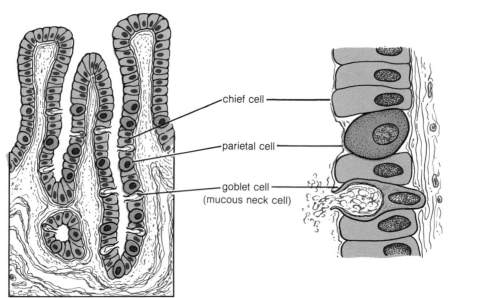

Figure 26-2 Three types of gastric gland cells make up the gastric glands that line the stomach.

- It helps to dissolve certain mineral salts found in the foods we consume.
- It destroys the bacteria and microorganisms that enter the stomach in our foods. This action is so efficient that partially digested food leaving the stomach is virtually sterile under normal conditions.

The enzyme *pepsin* acts only on proteins in the stomach. The optimum conditions for the functioning of pepsin are an acidic pH of 1.5 and a temperature of 37°C (body temperature of 98.6°F). Pepsin breaks down large protein molecules into intermediate-sized protein molecules called *proteoses* and *peptones*. As the proteose and peptone molecules are still too large for absorption into the bloodstream, they must pass into the small intestine for final digestion. It was once believed that human gastric juice contained a second protein-splitting enzyme called *rennin*. This was thought to act on the milk protein, *casein*.

It has since been proven that rennin does not exist in the adult stomach, and that pepsin can curdle milk. Rennin is, however, found in the stomach of infants and certain young mammals (such as calves). *Curdling* is the changing of the liquid casein into a solid form, or curd. It prepares milk for eventual digestion by other enzymes. It is of interest to note that rennin, obtained from calves' stomachs, is used commercially to curdle milk for cheeses and various milk desserts.

The action of the gastric juice is helped by the churning of the stomach walls. The semiliquid food which results is called *chyme*. When the chyme is ready to leave the stomach, a valve (the *pyloric sphincter*) at the lower end of the stomach opens from time to time. This allows the food to spurt on into the duodenum. The contraction and relaxation of smooth muscles in the walls of the alimentary tract move the food along the entire alimentary canal.

## Further Study and Discussion

- Identify each organ of the digestive system on a wall chart or a torso model.

- Add 5 mL of 0.5% hydrochloric acid to an equal volume of warm milk; note the reaction. Can this action be related to digestion?

- Explain why the stomach may change its shape before and after a meal.

- If laboratory facilities are available, place a small quantity of finely chopped, hardboiled eggwhite into each of four test tubes. To test tube #1, add 5 mL water; to test tube #2, add 5 mL 0.5% hydrochloric acid; to test tube #3, add a tiny amount of pepsin in 5 mL water; to test tube #4, add both a little pepsin and 5 mL of 0.5% hydrochloric acid. Place all the test tubes in an incubator overnight. Observe the results and complete the following table.

| TEST TUBE | CONTENTS | OBSERVATION | CONCLUSION |
|---|---|---|---|
| 1 | White of egg plus water | No change | Water alone does not dissolve protein |
| 2 | White of egg plus 0.5% hydrochloric acid | | |
| 3 | White of egg plus pepsin and water | | |
| 4 | White of egg plus pepsin plus 0.5% hydrochloric acid | | |

## Assignment

A. Briefly answer the following questions.

1. List the organs of the alimentary canal and the accessory organs of digestion (as discussed in the preceding unit).

   *Alimentary Canal*           *Accessory Organs*

2. Explain the importance of hydrochloric acid in the stomach.

3. Name the main enzyme found in gastric juice and explain its function.

B. Match each term in column I with its correct description in column II.

| Column I | Column II |
|---|---|
| _____ 1. chyme | a. semiliquid condition of food found in the stomach |
| _____ 2. esophagus | b. acts upon protein in the stomach |
| _____ 3. hydrochloric acid | c. oversecretion may cause peptic ulcer |
| _____ 4. pepsin | d. involuntary muscle action of alimentary canal |
| _____ 5. peristalsis | e. passageway to the stomach |
| _____ 6. pharynx | f. storage place for food |
| _____ 7. ptyalin | g. lower section of stomach |
| _____ 8. pylorus | h. passage where swallowing action takes place |
| _____ 9. stomach | i. enzyme which changes starch to sugar |
| _____ 10. fundus | j. upper portion of stomach |
|  | k. sphincter muscle at entrance to stomach |

C. List the parts of the stomach and the organs through which food enters and leaves the stomach. (Study figure 26-1 before answering.)

# Unit 27
# DIGESTION IN THE SMALL INTESTINE

## KEY WORDS

| | | |
|---|---|---|
| absorption | glycogen | liver |
| bile | intestinal juice | pancreatic juice |
| gallbladder | lacteal | villi |

## OBJECTIVES

- Describe how the small intestine prepares food for absorption
- Describe the digestive function of the liver
- Explain the digestive function of the gallbladder
- Define the Key Words relating to this unit of study

The small intestine is a coiled portion of the alimentary canal and is about twenty-five feet long and one inch in diameter. It contains thousands of small intestinal glands which produce intestinal juice. In addition to intestinal juice, bile from the liver and pancreatic juice from the pancreas are poured into the duodenum, the first part of the small intestine, see color plate 14.

Liver bile is needed for the digestion of fat. It breaks up the fat into small droplets upon which the digestive juices can act. Pancreatic juice contains enzymes that: (a) continue the digestion of protein started in the stomach, (b) act on starch, and (c) digest fat. The enzymes of the intestinal juice complete the digestion of proteins and carbohydrates. Therefore, the combined action of bile, pancreatic juice, and intestinal juice completes the process of breaking down food mass into substances which can be absorbed into the bloodstream.

Absorption is possible because the lining of the small intestine is not smooth. It is covered with millions of tiny projections called *villi*. Each microscopic villus contains a network of blood and lymph capillaries. The digested portion of the food passes through the villi into the bloodstream and on to the body cells. The undigestible portion passes on to the large intestine.

## THE LIVER AND GALLBLADDER

During the process of digestion, the liver, a large organ located just below the diaphragm on the right side, mainly acts on fat metabolism. It manufactures bile and passes it along to its storehouse, the gallbladder. When bile is needed for the digestion of fats, the gallbladder releases it through a duct into the duodenum of the small intestine.

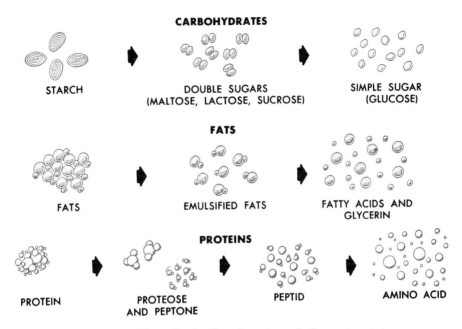

Figure 27-1  Phases in the digestion of starch, fat, and protein

In addition to manufacturing bile, the liver produces and stores glycogen (animal starch) from unused digested sugars. The liver also aids in removing certain waste products from the bloodstream, changing them into a form that can be excreted by the kidneys.

The bile contains mineral salts. If stored too long, these salts may crystallize and form gallstones either in the gallbladder or in the ducts through which the bile passes. Gallstones may keep the bile from reaching the small intestine.

Figure 27-1 shows how starch, fat, and protein are broken into simple forms and made ready for absorption.

## Further Study and Discussion

- Discuss reasons for giving hospital patients dilute glucose solution through their veins instead of a regular diet by mouth.

## Assignment

Match each term in column I with its description in column II.

| Column I | Column II |
|---|---|
| _____ 1. small intestine | a. substances which contain enzymes capable of acting on the digestion of proteins, starch and fats |
| _____ 2. end result of protein digestion | b. tiny projections in the small intestines which greatly increase absorption area |
| _____ 3. villi | c. receives the undigested portion of food at the end of the alimentary canal |
| _____ 4. pancreatic juice | d. region into which bile from the liver and pancreatic juice are poured |
| _____ 5. enzymes of intestinal juice | e. amino acids |
| _____ 6. duodenum | f. fatty acids and glycerin |
| _____ 7. liver bile | g. sucrose |
| _____ 8. large intestine | h. is about 25 feet long and one inch wide |
| _____ 9. glucose | i. emulsifies fat |
| _____ 10. usable products of fat metabolism | j. digestive juice for fat metabolism |
| | k. substance which may result from the breakdown of starch |
| | l. enzymes which complete digestion of proteins and carbohydrates |

# Unit 28
## THE LARGE INTESTINE

## KEY WORDS

anal sphincter
anus
appendicitis
ascending colon
bowel
cathartic
cecum
cellulose
colon
colonic stasis
defecation
descending colon
feces
hemorrhoids (piles)
ileocecal (colic) valve
laxative
mass peristalsis
nonpathogenic
rectal columns
rectum
sigmoid colon
tranverse colon
vermiform appendix

## OBJECTIVES

- Locate the large intestine
- Describe the functions of the large intestine
- List foods which aid the function of the colon
- Define the Key Words related to this unit of study

The large intestine, or *colon*, is 1.5 meters long (about 5 feet) and 2 inches in diameter (twice as wide as the small intestine). It is called the large intestine because of its larger lumen. The colon forms three sides of a square starting at the lower right hand corner of the abdominal cavity (up, across, and down).

## CECUM AND APPENDIX

The ileum empties its intestinal chyme into the side wall of the colon through an opening called the *colic* or *ileocecal valve*, figure 28-1. This valve permits passage of the intestinal chyme toward the colon, while preventing backsliding into the ileum. Located slightly below this valve, in the lower right portion of the abdomen, is a blind pouch which we call the *cecum*.

Just below the ileocecal valve, to the lower left of the cecum, is the *vermiform appendix*. The appendix is a finger-like projection protruding into the abdominal cavity. It has no digestive function. Because the appendix is a blind sac, it fills up easily, but drains quite slowly; substances can remain within the appendix for prolonged periods. Irritation of the lining of the appendix, which may be caused by hard or rough materials, can make it a suitable area for bacterial growth. This often leads to the painful inflammatory condition known as *appendicitis*.

## COLON

The colon continues upward, along the right side of the abdominal cavity, to the underside of the liver, forming the *ascending colon*. Then it veers abruptly to the left of

**184** SECTION 6 DIGESTION OF FOOD

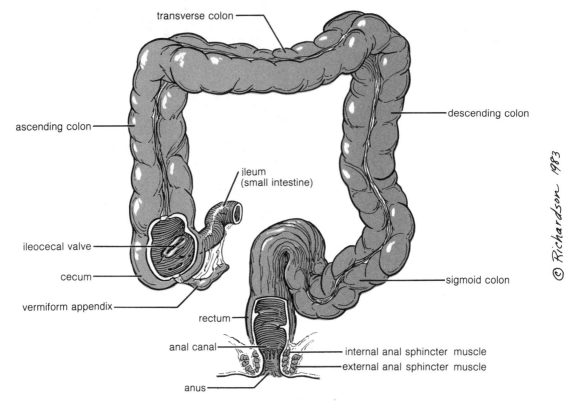

Figure 28-1 The large intestine

the abdominal cavity, near the level of the third lumbar vertebra, forming the *transverse colon*. The *descending colon* travels down the left side of the abdominal cavity. As the descending colon reaches the left iliac region, it enters the pelvis in an S-shaped bend. This section is known as the *sigmoid colon,* which extends some 7 or 8 inches as the *rectum*. The rectum opens exteriorly into the *anus*.

## ANAL CANAL

The anal canal is the last portion of the large intestine; its external opening is the anus. The anus is guarded by two *anal sphincter muscles*. One is an internal sphincter of smooth, involuntary muscle, the other an external sphincter of striated, voluntary muscle. Both of these remain contracted to close the anal opening until defecation takes place. The mucous membrane lining the anal canal is folded into vertical folds called *rectal columns*. Within each rectal column is an artery and a vein. The condition leading to inflammation or enlargement of the rectal column veins is known as *hemorrhoids,* or *piles*.

## FUNCTIONS OF THE LARGE INTESTINE

The large intestine is concerned with water absorption, bacterial action, fecal formation, and defecation. The purpose of these functions is to regulate the body's water balance while storing and excreting waste products of digestion.

## Water Absorption

The large intestine aids in the regulation of the body's water balance by absorbing large quantities of water back into the bloodstream. The water is drawn from the undigested food and indigestible material (like cellulose) that passes through the colon.

## Bacterial Action

A few hours following the birth of an infant, the lining of the colon starts to accumulate bacteria. These bacteria enter the body via ingested food and persist throughout the person's lifetime. A proportion of the bacteria are destroyed by the action of hydrochloric acid contained in gastric juice in the stomach. Those bacteria that survive are passed along to the colon, where they multiply rapidly, to form the bacterial population or flora, of the colon. The intestinal bacteria are harmless (*nonpathogenic*) to their host. They act upon the undigested food remains, turning them into acids, amines, gases, and other waste products. Some of these decomposed products are eventually excreted through the colon. Another benefit of the bacterial action is the synthesis (formation) of moderate amounts of B-complex vitamins and vitamin K (needed for blood clotting).

## Fecal Formation

Initially, the undigested or indigestible material in the colon contains a lot of water and is in a liquid state. Due to water absorption and bacterial action, it is subsequently converted into a semisolid form, called feces.

*Feces* consist of bacteria, waste products from the blood, acids, amines, inorganic salts, gases, mucus, and cellulose. The acids are acetic, butyric, and lactic. Amines that are produced are indole and skatole. Amines are waste products of amino acids. The gases are ammonia, carbon dioxide, hydrogen, hydrogen sulfide, and methane. The characteristically foul odor of feces derives from these substances — especially indole, skatole, and hydrogen sulfide.

Cellulose is the fibrous part of plants that humans are unable to digest. It contributes to the bulk of the feces. This bulk stimulates the muscular activity of the colon, resulting in defecation. Regular defecation (regularity) can be promoted by exercising daily and eating foods containing bulk, like cereals, fruits and vegetables. These foods supply the necessary roughage to initiate bowel movements.

## Defecation

Once approximately every 12 hours, the fecal material moves into the lower *bowel* (lower colon and rectum) by means of a series of long contractions called *mass peristalsis*. When the rectum becomes distended with the accumulation of feces, a defecation reflex is triggered. Nerve endings in the rectum are stimulated, and a nerve impulse is transmitted to the spinal cord. From the spinal cord, nerve impulses are sent to the colon, rectum, and internal anal sphincter. This causes the colon and rectal muscles to contract and the internal sphincter to relax, resulting in emptying of the bowels.

For defecation to occur, the external anal sphincter must also be relaxed. The external anal sphincter surrounds and guards the outer opening of the anus and is under conscious control. Due to this control, defecation can be prevented when inconvenient, despite the defecation reflex. However, if this urge is continually ignored, it lessens or disappears totally, resulting in constipation, or *colonic stasis*. Temporary relief from constipation may be obtained with the use of *laxatives* and *cathartics*. (A laxative is a substance that induces gentle bowel movement; a cathartic stimulates more vigorous movement, which may eventually reduce the bowel's muscle tone.)

## Further Study and Discussion

- Discuss what effect waste products could have on the general health were they not removed from the intestines.
- Discuss the special care needs of a person on bedrest based upon the information in this unit.
- Prepare a list of foods commonly recommended for persons suffering from constipation.

## Assignment

A. Briefly answer the following questions.

1. What causes constipation?

2. What is the function of the large intestine other than storage and elimination of wastes?

B. Read each statement carefully and determine if it is true or false. Encircle the letter *T* for true or *F* for false.

T   F   1. The large intestine is called the colon.

T   F   2. The large intestine is 20 feet long and 2 inches wide.

T   F   3. The cecum is located where the small intestine joins the large intestine.

T   F   4. The function of the appendix is unknown.

T   F   5. The large intestine stores and eliminates the waste products of digestion.

T   F   6. Regulation of water balance occurs in the large intestine because its lining absorbs water.

T   F   7. Constipation may be overcome by intensive and long periods of work and exercise.

T   F   8. Bulk foods such as cereals, fruits and vegetables may help avoid constipation.

T   F   9. The rectum is an extension of the descending colon.

T   F   10. The transverse colon lies between the ascending and the descending colon.

# Unit 29
# REPRESENTATIVE DISORDERS OF THE DIGESTIVE SYSTEM

## OBJECTIVES

- Identify common disorders which interfere with digestion
- Relate general treatment to these common disorders

It is well to know the common disorders of the digestive system and the general treatment of them. Some of these are briefly described in this unit.

## Stomatitis

*Stomatitis* is an inflammation of the soft tissues of the mouth cavity. Pain and salivation may occur also.

## Hiatal Hernia

*Hiatal hernia,* or rupture, occurs when the stomach protrudes above the diaphragm through the esophagus opening. Changes in the diet may relieve the heartburn; surgery is not usually required.

## Heartburn

So-called *"heartburn"* results from a backflow of the highly acidic gastric juice into the lower end of the esophagus. This irritates the lining of the esophagus, causing a burning sensation. Some individuals suffer chronic heartburn, occasioned by improper closure of the cardiac constrictor muscle at the junction of the esophagus and stomach. Temporary relief from this condition can usually be obtained by ingesting a solution of bicarbonate of soda. This is an alkaline, or basic substance that will neutralize the stomach's gastric juices.

## Gastritis

*Gastritis* is an acute or chronic inflammation of the mucous membrane lining the stomach. It may be caused by irritants such as highly spiced foods or some drugs. There are many forms of gastritis, including:

- Atrophic gastritis — a chronic form in which the mucous membrane of the stomach has atrophied.

- Corrosive gastritis — this is an acute form of gastritis caused by corrosive poisons.
- Infectious gastritis — acute gastritis, associated with infectious diseases such as measles and scarlet fever.

## Peptic Ulcers

*Peptic ulcers* are lesions which occur in either the stomach (gastric ulcers) or small intestine (duodenal ulcers). This condition affects approximately 1 of 10 adults in the United States, usually affecting 4 times as many men as women.

Increased psychological stress contributes to the development of peptic ulcers. The ulcers result from insufficient mucus secretion, and from oversecretion of gastric juice containing hydrochloric acid in the stomach. This process wears away the stomach's mucosal wall; it may even perforate the stomach or duodenum, leading to peritonitis and hemorrhage.

Most peptic ulcers are of the duodenal types. The pain accompanying a duodenal ulcer comes from the irritation of exposed nerves and muscle cells near the ulcer. Temporary relief from pain can be obtained by taking of alkali substances, milk, and certain foods. These substances can neutralize the hydrochloric acid in the gastric juice and delay emptying of the stomach.

The characteristic burning pain associated with a peptic ulcer appears 2-3 hours after eating. By then, the food within the stomach has passed along into the small intestine; only the acid from the stomach is entering the duodenum.

## Pyloric Stenosis

*Pyloric stenosis* is a narrowing of the pyloric sphincter at the lower end of the stomach. It is often found in infants. Projectile vomiting may result; surgery is often necessary.

## Gastroenteritis

*Gastroenteritis* is the inflammation of the mucous membrane lining of the stomach and intestinal tract. This is a common disorder of infants and may lead to severe diarrhea and dehydration.

## Infectious Hepatitis

*Infectious hepatitis* is a viral infection of the liver, often spread through contaminated water or food. Symptoms include chills, fever, malaise, gastrointestinal disturbances, and jaundice. The skin takes on a yellowish tinge due to the excess bile in the bloodstream.

## Cirrhosis

*Cirrhosis* is a chronic, progressive inflammatory disease of the liver, characterized by growth of connective tissue. It is commonly caused by lack of proper nutrients.

## Gallstones

Bile is normally stored in the gallbladder, from which it is released into the small intestine for fat emulsification; it can thus be concentrated 5-10 times by the resorption of water.

Sometimes collections of crystallized cholesterol form in the gallbladder. These are combined with bile salts and bile pigments to form *gallstones*. Gallstones can block the bile duct, causing pain and digestive disorders. In such cases, bile cannot flow into the small intestine to help in fat emulsification, digestion, and absorption. Most gallstones are small and may pass with undigested food. However, the larger and obstructive ones must be surgically removed.

## Cholecystitis

*Cholecystitis* is the inflammation of the lining of the gallbladder. The disorder may cause blockage of the cystic duct.

## Peritonitis

*Peritonitis* is a condition in which the serous membrane lining of the abdominal cavity is inflamed. Vomiting and pain are symptoms of this condition.

## Diarrhea

If the feces are passed along the colon too rapidly, insufficient water is reabsorbed, and the feces become watery. *Diarrhea* is characterized by loose, watery, and frequent bowel movements. It may result from irritation of the colon's lining by dysentery bacteria, poor diet, nervousness, toxic substances, or from irritants in food (as in prunes, which stimulate intestinal peristalsis).

Excessive water loss from chronic or severe diarrhea may lead to ulcerative colitis. This situation is caused by the rapid flow of digestive juices from the small intestine into the colon. Eventually the action of the digestive juices can lead to ulcers in the colon wall.

## Chronic Constipation

Feces eliminated through the rectum are normally in a semisolid state. When defecation is delayed, however, the colon absorbs excessive water from the feces rendering them dry and hard. When this occurs, defecation (or evacuation) becomes difficult.

For this reason, suppressing the need to defecate at normal times can lead to constipation. Constipation can also be caused by emotions such as anxiety, fear, or fright. Headaches and other symptoms that frequently accompany constipation result from the distension of the rectum, as opposed to toxins from the feces.

Treatment usually consists of eating proper foods, especially cereals, fruits and vegetables; drinking plenty of fluids; getting enough exercise; setting regular bowel habits; and avoiding tension as much as possible.

## Carcinoma

*Carcinoma,* or cancer, may occur in any part of the digestive tract. The term "cancer" is a general word applied to a disease characterized by an abnormal and uncontrolled growth of cells. The resulting mass, or tumor, can invade and destroy surrounding normal tissues. Cancer cells from the tumor can spread, or *metastasize,* through the blood or lymph to start new cancerous growths in other parts of the body.

## Stomach Cancer

The initial cancer cells that develop in the stomach quickly grow into masses of tissue known as tumors. Such tumors may be benign or malignant. A malignant growth, or tumor, is a cancer.

*Benign* tumors do not metastasize. They usually can be removed completely and are not likely to recur. A malignant stomach tumor, however, invades neighboring healthy stomach cells and organs.

Malignant stomach cancer cells can spread to other body parts, forming new growths or metastases. Even if the original tumor is surgically removed, the cancer may recur when malignant cancer cells have spread.

The initial symptoms of stomach cancer are much like those of other digestive disorders: heartburn, loss of appetite, persistent indigestion, slight nausea, a feeling of bloated discomfort after eating, and occasional mild stomach pain. Later symptoms include traces of blood in the feces, pain, weight loss, and vomiting.

Treatment involves surgical removal of the stomach tumor as soon as possible. Depending upon the size and the extent of growth of the tumor, part or all of the stomach may have to be removed.

If a stomach cancer has metastasized, surgical removal of the affected parts of neighboring organs, like the pancreas or spleen, is frequently required. But even with the most successful tumor excision, malignant cells can spread throughout the body via the blood. In such cases, *chemotherapy* (treatment with anticancer drugs) is prescribed. These drugs are administered into the bloodstream, circulating through the body to kill cancerous cells in any location of the body.

Radiation therapy plays a limited role in the treatment of stomach cancer. Very strong radiation doses are needed to kill the cancer cells, and they might also seriously damage neighboring healthy cells.

## Further Study and Discussion

- Discuss problems which result from relieving symptoms such as heartburn and diarrhea with over-the-counter drugs instead of seeking medical advice.

## Assignment

Match each disorder listed in column I with its description in column II.

| Column I | Column II |
| --- | --- |
| _____ 1. cirrhosis | a. frequent liquid bowel movements |
| _____ 2. gastroenteritis | b. chronic liver disease |
| _____ 3. peptic ulcers | c. protrusion of the stomach into the esophagus |
| _____ 4. hiatal hernia | d. viral infection of the liver |
| _____ 5. heartburn | e. inflammation of the abdominal cavity |
| _____ 6. diarrhea | f. obstruction of the hepatic duct |
| _____ 7. cholecystitis | g. inflammation of the stomach and intestinal lining |
| _____ 8. infectious hepatitis | h. inflammation of the gallbladder lining |
| _____ 9. pyloric stenosis | i. narrowing of sphincter in the stomach |
| _____ 10. peritonitis | j. cardiospasm |
| | k. lesions which may result from acid secretion |
| | l. common symptom characterized by a burning sensation |

# SELF-EVALUATION

## Section 6 DIGESTION OF FOOD

A. Label the structures which make up the digestive system.

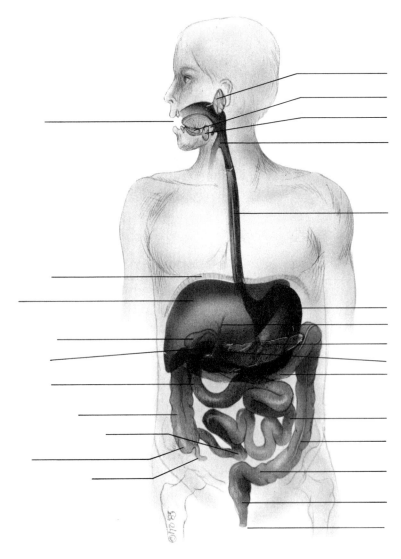

192

B. Complete the following statements.
   1. Substances which act upon foods to change them to simpler soluble forms are called _____.
   2. Teeth used for biting or cutting food are _____; those used for grinding are _____; and those used for tearing are _____.
   3. Digested food enters the bloodstream by passing through the _____ in the _____ intestine.
   4. The main functions of the large intestine are _____ and _____.
   5. The three juices which act upon food in the small intestine are ____, _____, and _____.

C. Select the item which best completes the statement.
   1. One part of the small intestine is the
      a. rectum
      b. duodenum
      c. pancreas
      d. appendix
   2. The appendix is attached to the
      a. duodenum
      b. rectum
      c. cecum
      d. pylorus
   3. Bile is secreted by the
      a. pancreas
      b. gallbladder
      c. stomach
      d. liver
   4. A substance which requires further digestion to break it down is
      a. fatty acid
      b. amino acid
      c. protein
      d. glycerol
   5. The salivary glands are situated
      a. in the small intestine
      b. in the pancreas
      c. in the mouth
      d. in the stomach

D. How does the blood differ in its composition before it enters the capillaries of the small intestine and after it leaves?

E. Explain the function of the liver and gallbladder in digestion.

# Section 7
# Elimination of Waste Materials

# Unit 30
# INTRODUCTION TO THE EXCRETORY SYSTEM

## KEY WORDS

dysuria
excretion
incontinence
metabolic wastes
micturition
retention

## OBJECTIVES

- Explain the function of the excretory organs
- List the parts of the body involved in elimination
- Relate the type of waste to the channel of excretion
- Define the Key Words related to this unit of study

Food is utilized through the process of digestion, absorption, and metabolism. These steps separate substances which can be digested from those which cannot. The blood and lymph transport products of digestion to the tissues where they are needed. After the cells of the tissues have used the food and oxygen needed for growth and repair, waste products formed by the process are taken away and excreted from the body. If they were left to accumulate in the body, the waste would act as poisons. The excretory organs eliminate the metabolic wastes and undigested food residue.

The channels through which elimination takes place include the kidneys, the skin, the intestines, and the lungs. The lungs, generally considered part of the respiratory system, serve an excretory function in that they give off carbon dioxide and water vapor during exhalation. The urinary system functions largely as an excretory agent of nitrogenous wastes, salts, and water, while the skin includes excretion of the dissolved wastes present in perspiration, mostly dissolved salts. The indigestible residue, water and bacteria are excreted by the intestines. The excretion of waste products is described and summarized in table 30-1.

Table 30-1  Elimination of Waste Products

| ORGAN | PRODUCT OF EXCRETION | PROCESS OF ELIMINATION |
|---|---|---|
| Lungs | carbon dioxide and water vapor | exhalation |
| Kidneys | nitrogenous wastes and salts dissolved in water to form urine | urination |
| Skin | dissolved salts | perspiration |
| Intestines | solid wastes and water | defecation |

## Further Study and Discussion

- Discuss what may happen to a person if the body does not regularly excrete waste materials.
- Discuss which part of the excretory system performs the most important function.

## Assignment

1. What is the function of the excretory organs?

2. What organs are involved in excretion?

3. How are waste products transported to the organs of excretion?

4. Name the excreted waste products of the body.

5. Explain the excretory function of the lungs.

# Unit 31
# URINARY SYSTEM

## KEY WORDS

adipose capsule
afferent arteriole
antidiuretic hormone
Bowman's capsule
distal convoluted tubule
efferent arteriole
fibrous capsule
glomerulus
hilum (hilus, pl.)
kidney
loop of Henle
nephric filtrate
nephron
peritoneum
peritubular capillaries
proximal convoluted tubule
renal column
renal corpuscle
renal cortex
renal fascia
renal medulla
renal papilla
renal pelvis
renal pyramids
reservoir
retroperitoneal
ureter
urethra
urinary bladder
urinary meatus

## OBJECTIVES

- List the organs which make up the urinary system
- Describe the way the kidneys excrete wastes from the body
- Explain how the kidneys regulate the water balance
- Define the Key Words related to this unit of study

The urinary system performs the main part of the excretory function in the body, see color plate 15A. The most important excretory organs are the *kidneys*. Their primary excretory function is removal of the nitrogenous waste products, the result of protein catabolism. If the kidneys fail to function properly, toxic wastes start to accumulate in the body. Toxic wastes accumulating in the cells cause them to "suffocate" and literally poison themselves.

The urinary system consists of two kidneys (that form the urine), two ureters, a bladder, and a urethra. Each kidney has a long, tubular *ureter* that carries urine to the *urinary bladder*. This is a temporary storage sac for urine, from which urine is excreted through the *urethra*.

## KIDNEYS

The kidneys are bean-shaped organs resting high against the dorsal wall of the abdominal cavity; they lie on either side of the vertebral column, between the peritoneum and the back muscles. Because the kidneys are located behind the peritoneum (rather than inside with the digestive organs), they are said to be *retroperitoneal.* They are positioned between the twelfth thoracic and the third lumbar vertebrae. The right kidney is situated slightly lower than the left, due to the large area occupied by the liver.

Each kidney and its blood vessels is enclosed within a mass of fat tissue called the *adipose capsule*. In turn, each kidney and adipose capsule is covered by a tough, fibrous tissue called the *renal fascia,* or *fibrous capsule.*

Since the kidneys are located behind the peritoneal cavity, they can be surgically approached from the back. They are cushioned and protected by the adipose capsule.

There is an indentation along the concave medial border of the kidney called the *hilum*. The hilum is a passageway for the lymph vessels, nerves, renal artery and vein, and the ureter. At the hilum the fibrous capsule continues downward, forming the outer layer of the ureter. Cutting the kidney in half lengthwise reveals its internal structure. The upper end of each ureter flares into a funnel-shaped structure known as the *renal pelvis*.

## Medulla and Cortex

The kidney is divided into two layers: an outer, granular layer called the *cortex,* and an inner, striated layer, the *medulla*. The medulla is red and consists of radially striated cones called the *renal pyramids*. The base of each renal pyramid faces the cortex, while its apex (*renal papilla*) empties into cuplike cavities called *calyces*. These, in turn, empty into the renal pelvis.

The cortex, reddish brown, is composed of millions of microscopic functional units of the kidney called *nephrons*. Cortical tissue is interspersed between renal pyramids, separating and supporting them. These interpyramidal cortical supports are the *renal columns*. The renal columns and the renal pyramids alternate with one another, see color plate 15B.

## NEPHRON

The nephron is the basic structural and functional unit of the kidney. Most of the nephron is located within the cortex, with only a small, tubular portion in the medulla. Over a million nephrons comprise each kidney.

A nephron begins with the *afferent arteriole,* which carries blood from the renal artery. The afferent arteriole enters a double-walled hollow capsule, the *Bowman's capsule.*[1] Within the capsule the afferent arteriole finely divides, forming a knotty ball called the *glomerulus* which contains some fifty separate capillaries. The combination of the Bowman's capsule and the glomerulus is known as the *renal corpuscle*. The Bowman's capsule sends off a highly convoluted (twisted) tubular branch referred to as the *proximal convoluted tubule*.

Both the renal corpuscle and the proximal convoluted tubule are located in the cortex. The proximal convoluted tubule descends into the medulla to form the *loop of Henle*. In figure 31-1, observe that the loop of Henle has a straight descending limb, a loop, and a straight ascending limb. When the ascending limb of Henle's loop returns to the cortex, it turns into the *distal convoluted tubule*. Eventually this convoluted tubule opens into a larger, straight vessel known as the *collecting tubule*. Several distal convoluted tubules join to form this single straight collecting tubule. The collecting tubule empties into the renal pelvis, then into the ureter.

As figure 31-1 shows, the walls of the renal tubules are surrounded by capillaries. After the *afferent* arteriole branches out to form the glomerulus, it leaves the Bowman's capsule as the *efferent* arteriole. The efferent arteriole branches to form the capillaries surrounding the renal tubules. All of these capillaries eventually join together to form a small branch of the renal vein which carries blood from the kidney.

---

1. Sir William Bowman (1816-1892), English anatomist and ophthalmologist

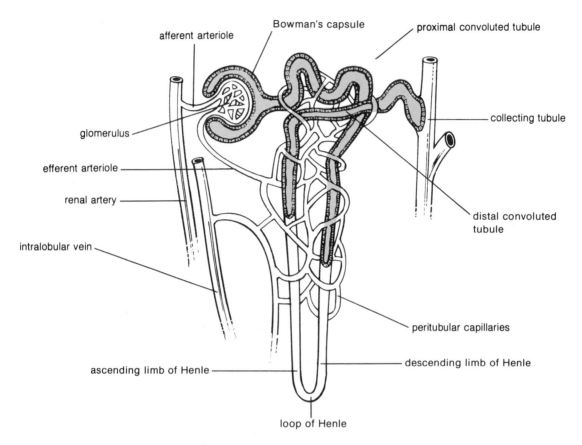

Figure 31-1 A nephron — the functional and structural unit of the kidney (from W. Schraer and R. Noelle, *A Learning Program for Biology,* Fairfield, New Jersey: Cebco Publishing Co.)

## URINE FORMATION IN THE NEPHRON

The kidney nephrons form urine by three processes: (a) filtration by the glomerulus, (b) reabsorption within the renal tubules, and (c) secretion by the tubular cells.

### Filtration

The first step in urine formation is filtration. In this process, blood from the renal artery enters the smaller afferent arteriole, which in turn enters the even smaller capillaries of the glomerulus. As the blood from the renal artery travels this course, the blood vessels grow narrower and narrower. This results in an increase in blood pressure. In most of the capillaries throughout the body, blood pressure is about 25 millimeters of mercury; in the glomerulus, it is between 60 and 90 millimeters.

This high pressure forces a plasma-like fluid to filter from the blood in the glomerulus into the Bowman's capsule. This fluid is called the *nephric filtrate*. It consists of water, glucose, amino acids, some salts, and urea. The nephric filtrate does not contain plasma proteins because they are too large to pass through the pores of the capillary membrane. The Bowman's capsule filters out 125 milliliters of fluid from the blood in a single minute. Thus in one hour, 7500 mL of filtrate leave the blood; this amounts to some 180,000 milliliters (180 liters) in a 24-hour period.

Obviously we cannot dispose of that much water each day. In fact, we only lose about a liter and a half per day because, as the nephric filtrate continues along the tubules, 99% of this water is reabsorbed back into the bloodstream.

## Reabsorption

This process includes the reabsorption of useful substances from the nephric filtrate within the renal tubules into the capillaries around the tubules (*peritubular capillaries*). These substances are water, glucose, amino acids, vitamins, bicarbonate ions ($HCO_3^-$), and the chloride salts of calcium, magnesium, sodium and potassium. Reabsorption starts in the proximal convoluted tubules; it continues through the Henle's loop, the distal convoluted tubules, and the collecting tubules.

The proximal tubules reabsorb approximately 80% of the water filtered out of the blood in the glomerulus (180 liters). Water thus absorbed through the proximal tubules constitutes *obligatory* water absorption. This process occurs by osmosis. Simultaneously, glucose, amino acids, vitamins, and some sodium ions are actively transported back into the blood.

In the distal convoluted tubules about 10 to 15 percent of water is reabsorbed into the bloodstream, depending upon the needs of the body. This type of water absorption is called facultative (optional) reabsorption. It is controlled by the *antidiuretic hormone (ADH)*, the hormone secreted from the posterior lobe of the pituitary gland, which is found at the base of the brain. ADH helps to maintain balance of body fluids by controlling the reabsorption of water in the nephron.

## Secretion

The process of secretion is the opposite of reabsorption. Some substances are actively secreted into the tubules. Secretion transports substances from the blood in the peritubular capillaries into the urine in the distal and collecting tubules. Substances secreted into the urine include ammonia, hydrogen ions ($H^+$), potassium ions ($K^+$), and some drugs. Hydrogen and potassium ions, as well as drugs, are secreted from the blood into the urine by active transport. Ammonia is secreted by diffusion.

## URETERS

Urine passes from the kidneys out of the collecting tubules into the renal pelvis, then down the ureter into the urinary bladder. There are two ureters (one from each kidney) carrying urine from the kidneys to the urinary bladder. They are long, narrow tubes, less than 1/4 inch wide and 10 to 12 inches long. Mucous membrane lines both renal pelves and the ureters. Beneath the mucous membrane lining of the ureters are smooth muscle fibers. When these muscles contract, peristalsis is initiated, pushing urine down the ureter into the urinary bladder.

## URINARY BLADDER

The *bladder*, a hollow muscular organ, made of elastic fibers and involuntary muscle, acts like a reservoir. It stores the urine until about one pint is accumulated. The bladder then becomes uncomfortable and must be emptied. Emptying the bladder, or voiding, takes place by muscular contractions of the bladder which are involuntary, although they can be controlled to some extent through the nervous system. Contraction of the bladder muscles forces the urine through a narrow canal, the *urethra,* which extends to the outside opening, the *urinary meatus.*

The kidneys have the potential to work harder than they actually do. Under ordinary circumstances, only a portion of the glomeruli are used. Should one kidney not function, or have to be removed, more glomeruli and tubules open up in the second kidney to assume the work of the nonfunctioning or missing kidney.

---

### Further Study and Discussion

- If laboratory facilities are available, obtain and examine several specimens of fresh, normal urine.

    a. What is the color of the specimen?

    b. Is it clear or cloudy?

    c. Is the urine acid, alkaline or neutral?

    > To test, dip blue litmus paper into the urine. If acid is present, it will turn red. Dip red litmus paper in. If urine is alkaline, it will turn paper blue. If neither paper changes color, the urine is neutral.

    d. What is the specific gravity of a specimen? To test, use a urinometer.

    e. Is albumin present?

    > To test for albumin, place 10 milliliters of urine in a test tube. Add 3 drops of dilute acetic acid (2%). Hold the tube at the bottom and apply heat to the upper level of urine. If a cloud appears in the heated portion, albumin is present.

    f. Is sugar (glucose) present?

    > To test for sugar, place 10 drops of urine in a Pyrex test tube. Add 5 milliliters of Benedict's solution. Mix thoroughly by shaking gently. Place in a water bath and boil for 3 minutes. As soon as the bubbling stops, interpret the test results as follows:

| COLOR | INDICATION |
|---|---|
| No change in color | 0 absent |
| Green | ± trace of sugar |
| Greenish-yellow | + one plus |
| Yellow | ++ two plus |
| Brown or brick red | +++ three plus |

- Using **Acetest** reagent tablets, examine the urine for acetone. Have the results and your interpretation checked by the instructor.

  Place the reagent tablet on a clean white sheet of paper. Place a drop of urine on the tablet. In 30 seconds, compare the resulting color with the color chart enclosed with the tablets. Record the result on the chart.

- Using **Clinitest** tablets and/or Clinistix reagent strips, test for sugar. Have the results and your interpretation checked by the instructor.

  Clinitest tablets: Place 5 drops of urine and 10 drops of water in a test tube. Add the Clinitest tablet. Observe the reaction. Then shake the test tube and compare the color of the solution with the color scale enclosed with the tablets. Record the result.

  Clinistix reagent strips: Dip the test end of the Clinistix in the urine and remove it. (Avoid contact with fingers or other objects because misleading results may occur.) If the moistened end turns blue, the result is *positive*. When sugar is present, the blue color will appear in less than one minute. Record the result.

- Using **Bumintest** reagent solution and/or Allritest tablets, test the urine for albumin. Have the results and your interpretation checked by the instructor.

  Bumintest: Dissolve 4 Bumintest reagent tablets in 30 milliliters of water in a test tube. (This makes a 5% solution). In another test tube, mix equal parts of urine and Bumintest solution and shake the tube gently. The amount of albumin is estimated by the degree of cloudiness (turbidity). Record the result.

  Allritest tablets: Place Allritest tablet on clean paper. Put one drop of urine on the tablet. When the urine has been absorbed, add 2 drops of water and allow the water to be absorbed before reading. Compare the color of the top of the tablet with the color photograph enclosed with each package of tablets. Record the result.

# Assignment

A. Match each term in column I with its description in column II.

| Column I | Column II |
|---|---|
| _____ 1. nephron | a. tubes which connect the kidney with the bladder |
| _____ 2. glomerulus | b. mass of capillaries |
| _____ 3. bladder | c. structure which absorbs wastes from the capillary mass |
| _____ 4. urethra | d. one of millions of tiny filtering units |
| _____ 5. ureter | e. returns blood to the inferior vena cava |
| _____ 6. ADH | f. hormone which regulates water reabsorption |
| _____ 7. tubules | g. contraction of bladder muscles |
| _____ 8. Bowman's capsule | h. canal which opens to the outside of the body |
| _____ 9. kidney | i. primarily acts as a reservoir |
| _____ 10. renal vein | j. allow urine to drain into the renal pelvis |
|  | k. bean-shaped organ |

B. Briefly answer the following questions.

1. How do the kidneys function in excretion?

2. Of what value is it to the physician to have an exact measure of the patient's intake and output?

3. What is another important function of the kidneys besides the elimination of waste?

# Unit 32
## THE SKIN

## KEY WORDS

albinism
callus
corn
dermis (corium or cutis vera)
end organs of Krause
end organs of Ruffini
epidermis (cuticle)
keratin
integument (integumentary system)
Meissner's corpuscles
melanin
melanocyte
Pacinian corpuscles
pili
sebaceous gland
stratum corneum
stratum germinativum
sudoriferous gland
ungues

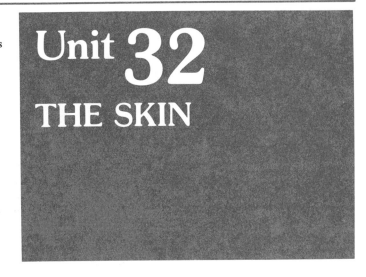

## OBJECTIVES

- Describe the functions of the skin
- Describe the structures found in the two skin layers
- Explain how the skin serves as a channel of excretion
- Describe the action of the sweat glands
- Define the Key Words related to this unit of study

The skin is our protective covering and is called the *integument* or *integumentary system*. It is tough, pliable, and multi-functional. Skin is first thought of as a covering for the underlying, deeper tissues, protecting them from dehydration, injury, and germ invasion. The skin also helps regulate body temperature by controlling the amount of heat loss. Evaporation of water from the skin, in the form of perspiration, helps rid the body of excess heat. Only a very small amount of waste is eliminated through the skin.

Since the skin is well supplied with nerves, it is sensitive to changes in its surrounding environment. These include changes in temperature (heat or cold), pain, pressure, and touch sensations.

The skin has tissues for the temporary storage of fat, glucose, water, and salts like sodium chloride. Most of these substances are later absorbed by the blood and transported to other parts of the body.

The skin is designed to screen out any harmful ultraviolet radiation contained in sunlight. It also absorbs certain drugs, and other chemical substances. The administration of drugs through the skin is called *inunction*. This absorptive quality can be harmful if insecticides, gas, or lead salts enter the body through the skin.

## STRUCTURE OF THE SKIN

The skin consists of two basic layers: (1) The *epidermis* or *cuticle* and (2) The *dermis corium* or *cutis vera* (true skin).

205

206 SECTION 7 ELIMINATION OF WASTE MATERIALS

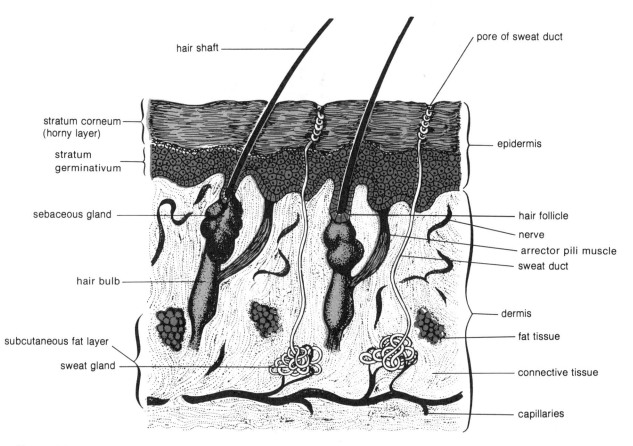

Figure 32-1 Cross section of the skin (from W. Schraer and R. Noelle, *A Learning Program for Biology,* Fairfield, New Jersey: Cebco Publishing Co.)

## Epidermis

The epidermis is made of two cellular layers known as the *stratum corneum* and the *stratum germinativum.* The cytoplasm of the cells making up the stratum corneum is replaced by a hard, nonliving protein substance called *keratin*. This keratin layer acts as a waterproof covering. Cells making up the stratum corneum are flattened and scalelike. They flake off from the constant friction of clothing, rubbing, and washing. For this reason the stratum corneum is sometimes called the *horny layer.* As the cells of the horny layer are flaked off, they are replaced by new cells from the lower stratum germinativum.

The stratum corneum forms the body's first line of defense against invading bacteria. Because it is slightly acidic, many kinds of organisms which come in contact with the stratum corneum are destroyed. The thickness of the horny layer varies in different parts of the body. It is thickest on the palms of the hands and on the soles of the feet due to constant friction. Sometimes the thickening develops outwardly in a concentrated area forming a *callus*. If the thickening grows inward, a *corn* may form.

The stratum germinativum is a very important epidermal layer for the growth of new cells in the epidermis depends upon the growth of cells in this layer. As new germinativum cells form, they push their way upward towards the epidermis. Eventually they become keratinized like the other epidermal cells within the horny layer.

Skin pigmentation is found in germinativum cells called *melanocytes*. Melanocytes contain a skin pigment called *melanin*. Melanin can be black, brown, or have a yellow tint, depending upon racial origin. The amount of melanin (and other skin pigments like carotene and hemoglobin) in the melanocytes determines the various shades of human skin color. Members of the Caucasian (white) race have a reduced amount of melanin in their melanocytes. Other races, on the other hand, possess a higher amount of melanin. Absence of pigments (other than hemoglobin) causes *albinism*. The skin of an albino has a pinkish tint. Basic skin coloring is inherited from our parents.

Environment is another factor which can modify skin coloring. For example, exposure to sunlight may result in a temporary increase in melanin within the melanocytes. This is the darkened, or tanned effect with which we are all familiar. Tanning is produced by the ultraviolet (UV) rays of sunlight. It should be noted that prolonged exposure to sunlight is unwise because it may lead to the development of skin cancers.

As seen in figure 32-1, the lower edge of the stratum germinativum is thrown into ridges. These ridges are known as the *papillae* of the skin. In the skin of the fingers, soles of the feet, and the palms of the hands, these papillae are quite pronounced. So much so, in fact, that they raise the skin into permanent ridges. These ridges are so arranged that they provide maximum resistance to slipping when grasping and holding objects. Thus they are also referred to as *friction ridges*. The ridges on the inner surfaces of the fingers create individual and characteristic fingerprint patterns used in identification. Newborn infants are also footprinted for means of identity.

## Dermis

The dermis, or *corium,* is the thicker, inner layer of the skin. It contains matted masses of connective tissue, strong fibrous tissue bands, elastic fibers (through which pass numerous blood vessels), lymphatics, nerve endings, muscles, hair follicles, oil and sweat glands, and fat cells. The thickness of the dermis varies over different parts of the body. It is, for instance, thicker over the soles of the feet and the palms of the hand. The skin covering the shoulders and back is thinner than that over the palms, but thicker than the skin over the abdomen and thorax.

There are many nerve receptors of different types in the dermal layer. The sensory nerves end in receptors which are sensitive to heat. They are called the *end organs of Ruffini*. Those sensitive to cold are the *end organs of Krause*. *Meissner's corpuscles* detect the sensation of touch, while the deeper lying *Pacinian corpuscles* detect pressure. There are also nerve endings to sense pain located under the epidermis and around the hair follicles. These pain receptors are especially numerous on the lower arm, breast, and forehead.

## APPENDAGES OF THE SKIN

The appendages of the skin include the hair, nails, the sudoriferous (sweat) glands, the sebaceous (oil) glands, and their ducts.

### Hair

Hairs (*pili*) are distributed over most of the surface area of the body. They are missing from the palms of hands, soles of feet, the glans penis, and the inner surfaces of the vaginal labia.

The length, thickness, type, and color of hair varies with the different body parts and different races. The hairs of the eyelids, for example, are extremely short, while hair from the scalp can grow to a considerable length. Facial and pubic hair are quite thick. The hair of Asian people is straight; that of Africans is very curly.

Microscopic examination reveals that a hair is composed of three layers: the outer *cuticle* layer, the *cortex,* and the inner *medulla.* The cuticle consists of a single layer of flat, scalelike, keratinized cells that overlap each other. The cortex is comprised of elongated, keratinized, nonliving cells. Hair pigment is located in the cortex, or in the medulla if one is present. In dark hair the cortex contains pigment granules; white hair indicates the presence of air.

A hair consists of a *root* and a *shaft.* The root is the part of the hair that is implanted in the skin. The shaft is that part which projects from the skin surface. The root is embedded in an inpocketing of the epidermis called the *hair follicle.* Toward the lower end of the hair follicle is a tuft of tissue called the *papilla,* which extends upward into the hair root. The papilla contains capillaries which nourish the hair follicle cells. This is important because the division of cells in the hair follicle gives rise to a new hair.

Attached to each hair follicle on the side toward which it slopes is a smooth muscle called the *arector pili muscle.* When the pili muscle is stimulated, as by a sudden chill, it contracts and causes the skin to pucker around the hair. This occasions the so-called "goosebumps" or "gooseflesh" condition. When this occurs, a small amount of oil is produced, due to pressure on the sebaceous glands.

### Nails

The nails (*ungues*) are hard structures covering the dorsal surfaces of the last phalanges of the fingers and toes. They are slightly convex on their upper surfaces and concave on their lower surfaces. A nail is formed in the *nail bed,* or matrix. Here the epidermal cells first appear as elongated cells. These then fuse together to form hard, keratinized plates. As long as a nail bed remains intact, a nail will always be formed. Occasionally, a nail is lost due to an injury or disease. However, if the nail bed is not damaged, a new nail will be produced.

### Sweat Glands

Actual excretion is a minor function of the skin; certain wastes dissolved in perspiration are removed. Perspiration is 99 percent water with only small quantities of salt and organic materials (waste products). Sweat, or *sudoriferous,* glands are distributed over the entire skin surface. They are present in large numbers under the arms, on the palms of the hands, soles of the feet, and forehead.

Sweat glands are tubular, with a coiled base and a tubelike duct which extends to form a pore in the skin, figure 32-1. Perspiration is excreted through the pores. Under the control of the nervous system, these glands may be activated by several factors including heat, pain, fever, and nervousness.

The amount of water lost through the skin is almost 500 milliliters a day. However, this varies according to the type of exercise and the environmental temperature. In profuse sweating a great deal of sodium chloride (salt) may be lost; it is vital to replace the loss of salt or water as soon as possible.

## Sebaceous Glands

The skin is protected by a thick, oily substance known as *sebum,* secreted by the *sebaceous glands.* Sebum lubricates the skin, keeping it soft and pliable.

## Further Study and Discussion

- Pour an equal amount of cold water in two beakers. Cover the water in one beaker with olive oil to prevent evaporation of water. Place them in a container of boiling water.

    a. In which beaker did the temperature rise more rapidly? Why?

    b. What conclusion may be drawn from this experiment in discussing heat loss and the regulation of body temperature in human beings?

- Take your temperature and record the result.

    a. By mouth

    b. By axilla

    c. By mouth, immediately after a cold drink

    d. By mouth, immediately after a hot drink

    Compare the various temperatures. Discuss the reasons for the differences.

## Assignment

1. In what way does the skin act as an organ of excretion?

2. What factors stimulate the sweat glands into activity?

3. Under normal circumstances, approximately how much water is lost daily through the skin?

4. What is perspiration?

5. What are the sudoriferous glands?

# Unit 33
# REPRESENTATIVE DISORDERS OF THE EXCRETORY SYSTEM

## KEY WORDS

acne vulgaris
anuria
carbuncles
catheter
cystitis
• cystitis cystica
• cystitis emphysematosa
dermatophytosis
eczema
furuncles
gangrene
hemodialysis
impetigo contagiosa
incontinence
micturition
necrosis
nephritis
oliguria
pruritis
psoriasis
pyelitis
pyelocystitis
pyelonephritis
ringworm
scabies
shingles (herpes zoster)
uremia
urethritis
urethrocystitis
urinary calculi (kidney stones)
urticaria (hives or nettlerash)

## OBJECTIVES

- List some disorders of the urinary tract
- Recognize some common skin disorders
- Describe the more common excretory system disorders
- Define the Key Words related to this unit of study

The normal functioning of the excretory system has been described. Some disorders of the intestines were covered in unit 29. There are also malfunctions and disorders of the urinary tract and the skin with which the health care provider should be familiar.

## DISORDERS OF THE URINARY TRACT

*Acute kidney failure* may be sudden in onset. Causes may be nephritis, shock, injury, bleeding, sudden heart failure, or poisoning. A common symptom is the absence of urine formation, which is termed *anuria*. Suppression of urine formation is dangerous; unless anuria is relieved, *uremia* will develop. Uremia is a toxic condition where the blood retains urinary waste products (like urea) because the kidneys fail to excrete them. Symptoms resulting from uremia are headaches, nausea, vomiting and in extreme cases, coma and death. When urine formation is diminished below the normal amount, a condition known as *oliguria* results.

An artificial kidney machine may be used to remove wastes normally excreted by a healthy kidney. This process is called *hemodialysis*.

*Cystitis* is the inflammation of the mucous membrane lining of the urinary bladder. It can be caused by bacterial infection or kidney inflammation which has spread to the bladder. Inflammation of the urinary bladder is likely to be more obvious than a kidney infection, since it usually leads to frequent and painful urination. Proper treatment involves administration of large doses of antibiotics like erythrocin to kill the bacteria causing the inflammation.

*Cystitis cystica* is a chronic inflammation of the urinary bladder. It is characterized by the presence of small, translucent mucus-containing cysts.

*Cystitis emphysematosa* is a type of cystitis where spaces in the urinary bladder wall are filled with gas. This condition is caused by bacterial fermentation of sugar in the urine, as sometimes occurs in diabetes.

*Incontinence* is also known as *involuntary micturition* (urination). Here an individual loses voluntary control over urination. Incontinence occurs in babies prior to proper toilet training since they lack control over the external sphincter muscle of the urethra. Thus, urination occurs whenever the bladder fills. Similarly, a person who has suffered a stroke, or one whose spinal cord has been severed, has no bladder control. Such patients may require an indwelling *catheter.* This is a tube inserted into the neck of the bladder through the urethra. It diverts and directs urine into a convenient bag or other receptacle outside the body.

*Pyelitis* is inflammation of the pelvis of the kidney, usually due to an infection.

*Pyelocystitis* is the inflammation of the renal pelvis and urinary bladder, due to an infection. Treatment involves the use of antibiotics to destroy the bacteria.

*Pyelonephritis* is the inflammation of the kidney tissue itself along with its renal pelvis. This condition generally results from an infection that has spread from the ureters. One of the symptoms is pyuria, the presence of pus in the urine. The usual course of treatment includes the administration of antibiotics.

*Nephritis* is an inflammation of the kidney, causing damage to the kidney tissue. The result of this condition is that the kidneys are unable to carry on the task of excretion in an efficient manner. (It is also called Bright's disease.)

*Acute nephritis* usually occurs in children and young people. It may be a complication of a communicable disease, especially scarlet fever. The streptococcus organism may be the cause. *Chronic nephritis* is a kidney condition which develops gradually in older people. Usually high blood pressure is also present. Hardening of the renal blood vessels may be the cause, or the glomeruli and tubules may have been destroyed over an extended period of time.

*Urethritis* is inflammation of the urethra.

*Urethrocystitis* is the inflammation of the urethra and the urinary bladder, usually due to an infection. It is alleviated with antibiotics such as the tetracyclines, kanamycin, and sulfisoxazole.

*Urinary calculi* is another name for kidney stones. Some of the materials contained in urine are only slightly soluble in water. Therefore when stagnation of urine occurs, the microscopic crystals of calcium phosphate may clump together to form kidney stones. These kidney stones slowly grow in diameter. They eventually fill the renal pelvis and obstruct urine flow in the ureter. Their presence may be extremely uncomfortable; "passing a stone" during urination can cause excruciating pain. If it is not possible to re-dissolve or pass kidney stones, they must be surgically removed. There are various causes for the formation of these stones, including extended immobility, dehydration, renal infection, or hyperparathyroidism.

*Tuberculosis of the kidney* is a destructive kidney disease caused by the tubercle bacillus, **Mycobacterium tuberculosis.** Treatment involves the use of streptomycin.

# REPRESENTATIVE DISORDERS OF THE SKIN

*Acne vulgaris* is a common and chronic disorder of the sebaceous glands. Its exact cause is presently unknown. Somehow the sebaceous glands secrete excessive oil, or sebum, which is deposited at the openings of the glands. Eventually this oily deposit becomes hard, or keratinized, plugging up the opening. This prevents the escape of the oily secretions, and the area becomes filled with leukocytes. The leukocytes cause the accumulation of pus. Acne occurs most often during adolescence and is marked by blackheads, cysts, pimples, and scarring. While there is no definite evidence to support the theory, it is believed that acne may be associated with the oversecretion of sex hormones.

*Athlete's foot (dermatophytosis)* is a contagious fungal infection. The fungus, or dermatophyte, infects the superficial skin layer and leads to skin eruptions. These eruptions are characterized by the formation of small blisters between the fingers and most often, the toes. Accompanied by cracking and scaling, this condition is usually contracted in public baths or showers. Treatment involves thorough cleansing and drying of the affected area. In addition, special antifungal antibiotics (such as Amphotericin B) are administered and antifungal powders are applied liberally.

*Eczema* is an acute, or chronic, noncontagious inflammatory skin disease. The skin becomes dry, red, itchy, and scaly. Various factors can lead to eczema. These may include diet, tight clothing, cosmetics, creams, medications, or soaps. Eczema caused by ingested drugs is known as *dermatitis medicamentosa*. That caused by sunlight or artificial ultraviolet radiation is called *dermatitis actinica*. Treatment consists of removal or avoidance of the causative agent, as well as application of topical medications containing hydrocortisone. The medication, however, only helps to alleviate the symptoms.

*Gangrene* is the necrosis, or death, of tissue cells resulting from the blockage of blood supply to an area, or from disease or direct injury. The prompt surgical removal of the gangrenous area is required in order to prevent the spread of necrosis to healthy neighboring tissues.

*Impetigo contagiosa* is an acute, inflammatory and contagious skin disease seen in babies and young children. It is caused by the staphylococcus or streptococcus organism. This disorder is characterized by the appearance of vesicles which rupture and develop distinct yellow crusts.

*Pruritus,* an intense itching stimulated by irritation of a peripheral sensory nerve, is a symptom rather than a disease. Diabetes mellitus, liver ailments, and thyroid disorders may cause pruritus. Thus treatment of the respective disease alleviates the pruritis.

*Psoriasis* is a chronic inflammatory skin disease characterized by the development of reddish patches which are covered with silvery-white scales. It affects the hands, feet and scalp (but not the face). The application of creams or lotions containing a specially refined coal tar compound may alleviate the discomfort and help control the disease.

*Ringworm* is a highly contagious fungal infection marked by raised, itchy, circular patches with crusts. It may occur upon the skin, scalp, and underneath the nails. Ringworm can be effectively treated with a drug called griseofulvin.

*Scabies,* also known as "seven-year itch," is a contagious skin disorder caused by a tiny insect parasite, **Sarcoptes scabies**. This affliction is characterized by multiple skin lesions, along with intense itching which occurs chiefly during the night. The itching is caused by the female insect burrowing beneath the epidermis to lay her eggs. Specific ointments, frequent baths, and change of clean clothing are prescribed.

*Urticaria,* hives, or nettle rash is a skin condition recognized by the appearance of intensely itching wheals or welts. These welts have an elevated, usually white, center with a surrounding pink area. They appear in clusters distributed over the entire body surface. The welts last about a day or two. Urticaria is generally a response to an allergy, such as an ingested drug or foods like citrus fruits, chocolate, fish, eggs, shellfish, strawberries, and tomatoes. Complete avoidance and elimination of the causative factor(s) alleviate the problem.

*Furuncles* are boils which are usually the result of staphylococcus infections in the hair follicles.

*Carbuncles* are hard, round, deeply-embedded and painful abscesses of the subcutaneous skin tissue. A carbuncle is much larger than a boil; its flat surface oozes pus from multiple points on its surface. Fever usually accompanies the appearance of carbuncles. When they eventually slough away, they leave a scarred cavity behind. Treatment may require incision, drainage of pus, and use of antibiotics.

*Shingles (herpes zoster)* is a skin eruption thought to be due to a virus infection of the nerve endings. It is commonly seen on the chest or abdomen, accompanied by severe pain known as herpetic neuralgia. The condition is especially serious in elderly or debilitated persons. The affected area must be treated by protecting it from air and from the irritation of clothing.

## *Further Study and Discussion*

- Explain the value of urinalysis in diagnosing cystitis or pyelitis.
- Inspect the skin on various areas of the body. Look for the skin pores, skin irritations, pimples, scaliness, and dryness. Discuss methods of proper cleansing of the skin; nutrition to improve the skin; the effect of sunshine on the skin; and sensitivity of skin to cosmetics.
- Visit a hemodialysis center. Ask for permission to read a patient's history. Write a short report to present to the class.
- Invite a staff member from a kidney transplant center to give a presentation to the class.

## *Assignment*

1. What may cause cystitis or pyelitis?

UNIT 33  REPRESENTATIVE DISORDERS OF THE EXCRETORY SYSTEM   215

2. What is a frequent result of nephritis?

3. How are kidney stones formed?

4. What is the value of hemodialysis to the patient whose kidneys are unable to remove waste products from the blood?

# SELF-EVALUATION

## Section 7
## ELIMINATION OF WASTE MATERIALS

A. Match each description in column I with the correct term in column II.

| Column I | Column II |
|---|---|
| _____ 1. waste product eliminated through lungs | a. anuria |
| _____ 2. blood filter | b. calculi |
| _____ 3. stones in the kidney | c. carbon dioxide |
| _____ 4. water and nitrogenous wastes | d. cystitis |
| _____ 5. inflammation of the mucous membrane lining the bladder | e. kidneys |
| | f. nephritis |
| _____ 6. kidney malfunction preventing elimination of metabolic wastes | g. perspiration |
| | h. uremia |
| _____ 7. helps regulate body temperature | i. ureter |
| _____ 8. urinary duct | j. urine |
| | k. urinometer |

B. Label the parts indicated on the diagram.

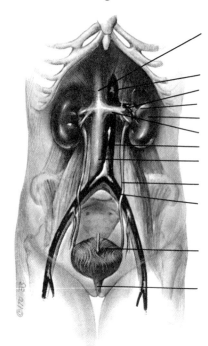

C. Match each term in column I with its correct description in column II.

| | Column I | Column II |
|---|---|---|
| _____ | 1. acne vulgaris | a. skin disorder of adolescence, marked by pimples, blackheads, cysts, scars |
| _____ | 2. pruritus | b. chronic skin disease characterized by red areas covered with silvery patches |
| _____ | 3. eczema | c. "7-year itch" caused by tiny parasites which bore under outer layer of skin |
| _____ | 4. impetigo contagiosa | d. allergic condition sometimes due to diet, soap, or creams |
| _____ | 5. psoriasis | e. contagious skin disease of babies or young children, caused by the streptococcus or staphylococcus organism |
| _____ | 6. ringworm | f. itching of the skin, may be due to diabetes mellitus, liver, or thyroid disorders |
| _____ | 7. shingles | g. contagious fungus infection with red circular crusty patches on skin or in hair |
| _____ | 8. urticaria | h. sudden-appearing, raised pink areas which itch and sting; commonly called hives |
| _____ | 9. scabies | i. herpes zoster, skin eruption due to viral infection of nerve endings |
| _____ | 10. carbuncle | j. deep abscess requiring incision, drainage, and antibiotics |
| | | k. death of tissue cells |

# Section 8
# Human Reproduction

# Unit 34

# INTRODUCTION TO THE REPRODUCTIVE SYSTEM

## KEY WORDS

| | | |
|---|---|---|
| autosome | gamete | ovaries |
| cleavage | gametogenesis | somatic cell |
| coitus | gonads | sperm |
| corona radiata | hyaluronic acid | spermatogenesis |
| deoxyribonucleic acid | hyaluronidase | testes |
| fertilization (internal) | meiosis | zona pellucida |
| | oogenesis | zygote |
| | ova | |

## OBJECTIVES

- Contrast reproduction of simple cells and more complex forms of life
- Explain the process of fertilization
- Describe how physical traits are determined
- Define the Key Words related to this unit of study

All living organisms, whether unicellular or multicellular, small or large, must reproduce in order to continue their species. Humans and most multicellular animals reproduce new members of their species by sexual reproduction.

Specialized sex cells (*gametes*) must be produced by the gonads of both male and female sex organs before sexual reproduction can take place. The female gonads, called the *ovaries,* produce egg cells (*ova*). The male gonads, the *testes,* produce sperm. The formation of gametes within the gonads is known as *gametogenesis,* or *meiosis.* In the female, the specific meiotic process is called *oogenesis;* in the male, *spermatogenesis.*

In humans, the *somatic* (body) cells, including skin, fat, muscle, nerve, bone cells, etc. contain 46 chromosomes in the nucleus. Forty-four of these are *autosomes* (nonsex chromosomes). The remaining two are sex chromosomes. Each chromosome has a partner of the same size and shape so that they can be paired, figure 34-1. In the female, the somatic cells contain 22 pairs of autosomes, and a single pair of sex chromosomes (both are $X$ chromosomes). In the male, the combination is also 22 autosomal pairs and a single pair of sex chromosomes. However, the male sex chromosomal pair consists of an $X$ and a $Y$ chromosome.

Oogenesis and spermatogenesis reduces the chromosome number of 46 to 23 in the gametes. All multicellular organisms start from a fusion of two gametes: the sperm (spermatozoan) from the male, and the ovum from the female. Figure 34-2 shows the structure of a spermatozoan and an ovum. During sexual intercourse, or *coitus,* sperm from the testes is deposited into the female

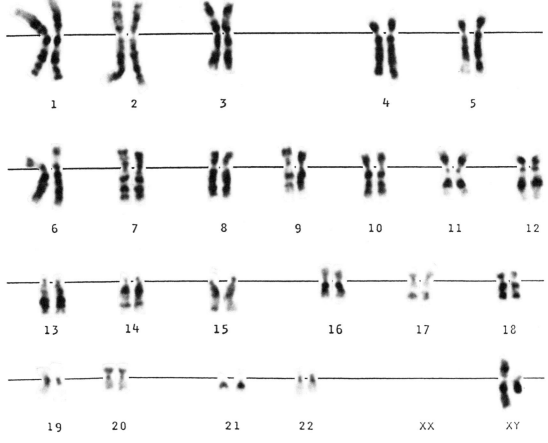

Figure 34-1 Karotype of human from a male somatic cell. A karotype is the arrangement of chromosome pairs according to shape and size.

vagina. Spermatozoa entering the female reproductive tract live for only a day or two at the most, though they may remain in the tract up to two weeks before degenerating. Approximately 100 million spermatozoa are contained in 1 milliliter (1cc) of ejaculated seminal fluid. They are fairly uniform in shape and size. If the count is less than 20 million per milliliter, the male is considered to be sterile. These millions of sperm cells swim towards the ovum that has been released from the ovary. The large quantity of sperm is necessary because a great number are destroyed before they even approach the ovum. Many die from the acidity of the secretions in the male urethra or the vagina. Some cannot withstand the high temperature of the female abdomen, while others lack the propulsion ability to progress from the vagina to the upper uterine (fallopian) tube.

In order for a sperm to penetrate and fertilize an ovum, the *corona radiata* must first be penetrated. This is the layer of epithelial cells surrounding the zona pellucida, figure 34-2. Eventually, only one sperm cell penetrates and fertilizes an ovum. To

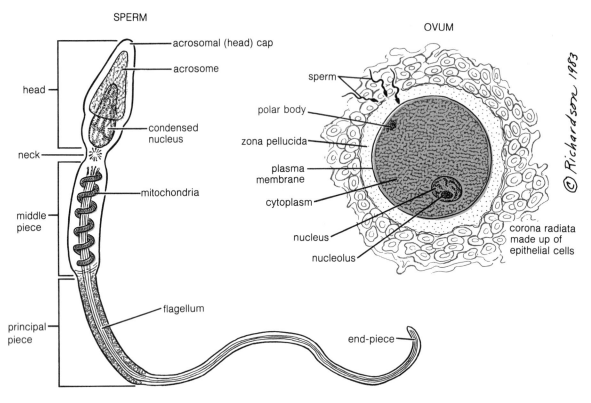

Figure 34-2 Diagrammatic representation of a human sperm and ovum

accomplish this successfully, the sperm head produces an enzyme called *hyaluronidase.* Hyaluronidase acts upon *hyaluronic acid,* a chemical substance that holds together the epithelial cells of the corona radiata. As a result of the action of the hyaluronidase, the epithelial cells fall away from the ovum. This exposes an area of the plasma membrane for sperm penetration. Figure 34-3 illustrates the fertilization of an ovum by a sperm.

*True fertilization* occurs when the sperm nucleus combines with the egg nucleus to form a fertilized egg cell, or *zygote.* The type of fertilization that occurs in humans is referred to as *internal fertilization;* fertilization takes place within the female's body.

Fertilization restores the full complement of 46 chromosomes possessed by every human cell, each parent contributing one chromosome to each of the 23 pairs.

A substance called *deoxyribonucleic acid* (DNA) is found in the chromosomes. It contains the genetic code that is replicated and passed on to each cell as the zygote divides and redivides to form the embryo. The early process whereby the zygote repeatedly divides to form an early embryo is known as *cleavage.* After early cleavage, actual embryonic development occurs until the fetus is completely formed.

All of the inherited traits possessed by the offspring are established at the time of

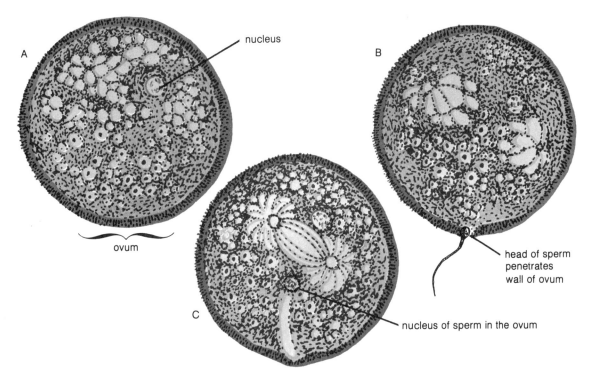

Figure 34-3 Fertilization

fertilization. This is a point to remember when working with parents. A young mother-to-be may hope that her baby will be a girl with curly hair, or a prospective father may insist that he wants a son. The health care provider can assure them that the sex, and physical characteristics such as eye color and curly hair, are determined at the time of fertilization. These traits cannot be altered by wishful thinking. The sex chromosomes of the male parent determine the sex of the child but other characteristics are a combination of both parents.

## Further Study and Discussion

- Discuss recessive and dominant heredity traits. Give examples of each.
- Find out if mental illness, tuberculosis, or cancer is inherited. Research and submit a paper on one of these topics.

## Assignment

1. When is the sex of the offspring determined?

2. What are chromosomes?

3. How many chromosomes are present in each body cell of the newborn child?

4. Where do these chromosomes come from?

5. What part does DNA play in hereditary characteristics?

# Unit 35
# THE ORGANS OF REPRODUCTION

## KEY WORDS

| | | |
|---|---|---|
| cervix | fimbrae | nongravid |
| corpus luteum | fundus | ova |
| cytogenic | graafian follicle | ovulation |
| ectopic pregnancy | infundibulum | penis |
| endocrinic | internal orifice (os) | progesterone |
| endometrium | of the uterus | puberty |
| estrogen | maturation | serosa of the uterus |
| external os of cervix | menarche | sperm |
| fallopian tube | menopause | testes |
| (oviduct) | myometrium | vagina |

## OBJECTIVES

- Identify the organs of the female reproductive system
- Identify the organs of the male reproductive system
- Describe the functions of the reproductive organs
- Define the Key Words relating to this unit of study

The function of the reproductive system is to provide for continuity of the species. In the human, the female reproductive system is composed of two ovaries, two fallopian tubes, the uterus, and the vagina. The male reproductive system is made up of two testes, seminal ducts, glands and the penis. The principal male organs are located outside the body in contrast to the female organs which are largely located within the body.

## FEMALE REPRODUCTIVE SYSTEM

As shown in color plate 16, the female reproductive system consists of two ovaries, two fallopian tubes, the uterus, and the vagina. Accessory organs are the breasts.

### The Ovaries

The ovaries are the primary sex organs of the female. They are located on either side of the pelvis, lateral to the uterus, in the lower part of the abdominal cavity. Each ovary is about the shape and size of a large almond, measuring about 3 centimeters long and from 1.5 to 3 centimeters wide. An *ovarian ligament,* a short fibrous cord within the *broad ligament,* attaches each ovary to the upper lateral part of the uterus.

Ovaries perform two functions. They produce the female gametes, or *ova,* and the female sex hormones, *estrogen* and *progesterone.* Thus the ovaries are said to be both *cytogenic* (cell-producing) and *endocrinic* (hormone-producing). Table 35-1 outlines the functions of the female sex hormones.

Each ovary contains thousands of microscopic hollow sacs called *graafian follicles* in varying stages of development. An ovum slowly develops inside each follicle. The process of development from an immature ova to a functional and mature ova inside the graafian follicle is called *maturation.* (In

**Table 35-1 The Functions of Estrogen and Progesterone**

| HORMONE | FUNCTION |
|---|---|
| Estrogen | 1. Repairs and thickens the uterine lining by stimulating the production of new epithelial cells<br>2. Develops and maintains the female secondary sex characteristics (breast development, feminine body contour, etc.) |
| Progesterone | 1. Thickens the uterine lining so it can receive the developing embryo (provided estrogen has acted previously upon the uterine lining)<br>2. Decreases uterine contractions during pregnancy |

addition, the graafian follicle produces the hormone, estrogen.)

Usually a single follicle matures every twenty-eight days throughout the reproductive years of a woman. The reproductive years begin at the time of *puberty* and the *menarche* (initial menstrual discharge of blood). This usually occurs between ages 9 and 17 (the average is 12.5 years). The reproductive years end at the time of *menopause,* the cessation of menstruation. This usually occurs when the woman is about 45 to 50 years of age.

Occasionally two or more follicles may mature, thereby releasing more than one ova. As the follicle enlarges, it migrates to the outside surface of the ovary and breaks open, releasing the ovum from the ovary. This process is called *ovulation;* it occurs about two weeks before the menstrual period begins. However, the time of ovulation may vary depending on emotional and physical health, state of mind, and age. During a woman's reproductive years, she produces about 400 ova. The two ovaries alternate in the maturation and ovulation of an ova.

When an ovum is released from the ovary, it is about 0.09 millimeter in diameter and contains a very large nucleus.

The ovum also consists of cytoplasm and some yolk. This yolk is the initial food source for the growth of the early embryo. After ovulation, the ovum travels down one of the *fallopian tubes,* or *oviducts.* Fertilization of the ovum takes place only in the upper third of the oviduct. Therefore, the time of fertilization is limited to a day or two following ovulation. The ovum begins to deteriorate as it slowly travels down towards the lower part of the oviduct. If fertilization has not taken place, the ovum passes out of the body during menstruation.

The development of the follicle and release of the ovum occur under the influence of two hormones produced in the anterior lobe of the pituitary gland; they are the follicle-stimulating hormone (FSH) and the luteinizing hormone (LH). The follicle-stimulating hormone, FSH, also promotes the secretion of estrogen by the ovary. Estrogen promotes the rapid growth of the uterine lining (the *endometrium*) in preparation for possible implantation of a fertilized ovum.

Following ovulation the ruptured follicle enlarges, takes on a yellow fatty substance, and becomes the *corpus luteum* (yellow body). The corpus luteum secretes another ovarian hormone, progesterone, which functions to maintain the growth of the uterine lining. If the egg is not fertilized, the corpus luteum degenerates, progesterone production stops, and the thickened glandular endometrium sloughs off. The tiny blood vessels that supply the endometrium are ruptured, producing the characteristic blood flow of menstruation. Following menstruation the endometrium heals and then starts thickening again, marking the beginning of the new menstrual cycle.

UNIT 35 THE ORGANS OF REPRODUCTION  227

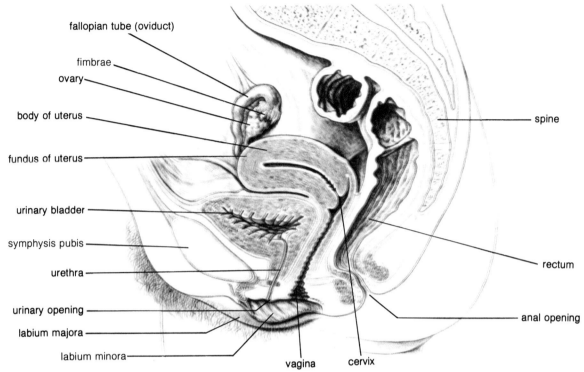

Figure 35-1 Longitudinal section of the female reproductive system (From *Human Anatomy and Physiology* by Joan G. Creager. © 1983 by Wadsworth Inc. Reprinted by permission of Wadsworth Publishing Company, Belmont, California 94002.)

## Fallopian Tubes

The fallopian tubes, about 10 centimeters (4 inches) long, are not attached to the ovaries, figure 35-1 and color plate 16. The outer end of each oviduct curves over the top edge of each ovary and opens into the abdominal cavity. This position of the oviduct, nearest the ovary, is the *infundibulum*. Since the infundibulum is not attached directly to the ovary, it is possible for an ovum to accidentally slip into the abdominal cavity and be fertilized there. Such a condition is known as "ectopic pregnancy" (developing outside the uterine cavity). In addition, an infection occurring in the infundibulum and ovary can spread to the abdominal cavity.

The area of the infundibulum over the ovary is surrounded by a number of fringelike folds called *fimbrae*. Each oviduct is lined with mucous membrane, smooth muscle, and ciliated epithelium. The combined action of the peristaltic contractions of the smooth muscles and the beating of the cilia helps to propel the ova down the oviduct into the uterus.

## Uterus

The *uterus* is a hollow, thick-walled, pear-shaped, and highly muscular organ. The *nongravid* (non-pregnant) uterus measures about 7.5 centimeters in length, 5 cm wide and 2.75 cm thick. This is about three inches long, two inches wide and about one inch thick. Its uterine cavity is extremely small and narrow. During pregnancy, however, the uterine cavity greatly expands in order to accommodate the growing embryo and a large amount of fluid.

The uterus is divided into three parts: (1) the *fundus,* the bulging, rounded upper part above the entrance of the two oviducts into the uterus; (2) the *body,* or middle portion; and (3) the *cervix,* or cylindrical, lower narrow portion that extends into the vagina. There is a short, cervical canal that extends from the lower uterine cavity (*internal orifice,* or *os of the uterus*) to the *external os* at the end of the cervix. The uterine wall is comprised of three layers:

- the outer *serous* layer
- an extremely thick, smooth, muscular middle layer, the *myometrium*
- an inner mucous layer, the *endometrium*

The endometrium, which lines the oviducts and the vagina, is also lined with ciliated epithelial cells, numerous uterine glands, and many capillaries.

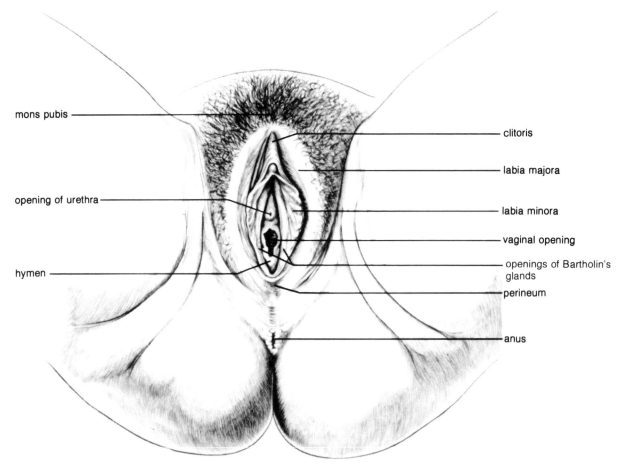

Figure 35-2 External female genitalia (From *Human Anatomy and Physiology* by Joan G. Creager. © 1983 by Wadsworth, Inc. Reprinted by permission of Wadsworth Publishing Company, Belmont, California 94002.)

During development of the embryo-fetus, the uterus gradually rises until the top part is high in the abdominal cavity, pushing on the diaphragm. This may cause the expectant mother some difficulty in breathing during the late stages of pregnancy.

## Vagina

The *vagina* is the short canal which extends from the cervix of the uterus to the vulva. It is muscular tissue which receives the sperm during sexual intercourse and stretches to assist in delivery during childbirth.

## Breasts

The breasts are accessory organs to the female reproductive system. They are composed of numerous lobes arranged in a circular formation. Clusters of secreting cells surround tiny ducts. A single duct extends from each lobe to an opening in the nipple. The *areola,* the darker area which surrounds the nipple, changes to a brownish color during pregnancy. *Prolactin* from the anterior lobe of the pituitary gland stimulates the mammary glands to secrete milk following childbirth.

# MALE REPRODUCTIVE SYSTEM

The male reproductive organs consist of the testes, a system of ducts for transporting sperm from the testes, several glandular structures, and the penis.

The *scrotum* is an external sac which contains two ovoid bodies called testes. They are the primary sex organs of the male and produce male *sperm,* and the male sex hormone, *testosterone.* This hormone influences the development of the secondary sexual characteristics of the male: deep voice; facial, pubic, and axillary hair; the typical male shape of wide shoulders and narrow hips. Testosterone is also necessary for the normal development of the secondary male reproductive organs: *epididymis, vas deferens, seminal vesicles, prostate gland,* and *penis,* figure 35-3.

The *gonadotrophic* hormones from the anterior lobe of the pituitary gland stimulate the testes to activity. Certain interstitial cells of the testes produce the testosterone. The process of sperm production begins shortly after puberty and continues until old age. Each testis produces millions of sperm, formed in the seminiferous tubules. From the testes the sperm passes through a system of ducts consisting of the epididymis, the vas deferens, the ejaculatory duct, and the urethra. Starting in the epididymis, certain secretions are added to the sperm along the route of travel and make up the *semen.*

In the male, the urethra serves a dual purpose as the outlet for the reproductive tract and the urinary tract. The prostate gland, one of the accessory organs which produces a secretion that forms part of the semen, sometimes causes trouble in later life because it may enlarge in older men. This enlargement presses against the urethra, sometimes making urinating impossible. Simple surgery (transurethral resection) can remedy this condition if performed in time. Otherwise, abdominal surgery may be required.

The *bulbourethral glands,* also known as Cowper's glands, are located on either side of the urethra, below the prostate gland. They add an alkaline secretion to the semen which helps the sperm to live longer within the acid medium of the female reproductive tract.

**230** SECTION 8 HUMAN REPRODUCTION

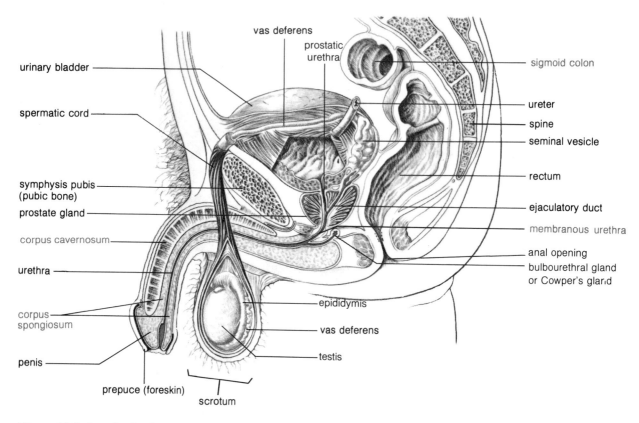

Figure 35-3 Longitudinal section of the male reproductive system (From *Human Anatomy and Physiology* by Joan G. Creager. © 1983 by Wadsworth, Inc. Reprinted by permission of Wadsworth Publishing Company, Belmont, California 94002.)

The external organs are the scrotum and the penis. Internally, the scrotum is divided into two sacs each containing a testis, epididymis, and lower part of the vas deferens. The penis contains erectile tissue which becomes enlarged and rigid during intercourse. Loose-fitting skin, called the *foreskin* or *prepuce,* covers the penis. The foreskin can be removed in a simple operation known as *circumcision.*

## Further Study and Discussion

- Discuss the relationship between ovulation, menstruation, and fertilization.
- Visit a prenatal clinic in your community. What do the examinations of a pregnant woman include on her first visit? On return visits?

- Discuss how oral and intrauterine contraceptives prevent pregnancy.
- Two blood tests which have been used to detect early pregnancy are the A-Z test and HCG test. How do they differ? What do the letters stand for?
- Investigate the value of using microsurgery techniques in ectopic or tubal pregnancies.

## Assignment

A. Select the letter of the item which most correctly completes the statement.

1. One of the male hormones is
   a. progesterone
   b. luteinizing hormone
   c. follicle-stimulating hormone
   d. testosterone

2. Ovulation usually occurs
   a. the day before the menstrual period begins
   b. one week before the menstrual period begins
   c. three weeks before the menstrual period begins
   d. two weeks before the menstrual period begins

3. The ovaries contain
   a. thirty graafian follicles
   b. thousands of graafian follicles
   c. hundreds of graafian follicles
   d. six graafian follicles

4. The development of the follicle and release of the ovum are under the influence of
   a. the follicle-stimulating hormone and the luteinizing hormone
   b. estrogen and corpus luteum
   c. progesterone and the follicle-stimulating hormone
   d. estrogen and the luteinizing hormone

5. Which one of the following statements is *not* correct?
   a. The fallopian tubes are about four inches long.
   b. The fallopian tubes serve as ducts for the ovum on its way to the uterus.
   c. The fallopian tubes are also called oviducts.
   d. The fallopian tubes are attached to the ovaries.

B. Match each term in column I with its correct description in column II.

|  | Column I | Column II |
|---|---|---|
| _____ | 1. scrotum | a. secondary sex characteristics |
| _____ | 2. testosterone | b. external sac which holds the testes |
| _____ | 3. facial and pubic hair | c. excreted from the pituitary gland |
| _____ | 4. epididymis and penis | d. formed in the seminiferous tubules |
| _____ | 5. gonadotrophic hormone | e. male gamete |
|  |  | f. secondary reproductive organs |
|  |  | g. male hormone produced in the testes |

# Unit 36
# REPRESENTATIVE DISORDERS OF THE REPRODUCTIVE SYSTEM

## KEY WORDS

benign
biopsy
carcinoma
dysmenorrhea
endocervicitis
epididymitis
fibroma
hysterectomy
leukorrhea
mastectomy
metastasis
orchitis
palpate
pap smear
prostatectomy
prostatitis
retroversion of the uterus
salpingitis

## OBJECTIVES

- List some common disorders of the reproductive system
- Identify symptoms of some common disorders
- Define the Key Words related to this unit of study

Persons who are involved in health care should be familiar with the names and symptoms of the more common disorders of the reproductive system. Some representative disorders and their symptoms are briefly described in this unit.

## FEMALE REPRODUCTIVE SYSTEM

A *carcinoma* is a malignant growth. Cancer cells multiply widely and invade normal cells and tissues. Carcinomas can develop in the breasts or uterus. As in other forms of cancer, early detection and treatment is vital to the patient's life. Breast carcinomas may sometimes be detected by self-examination. Periodically a woman should *palpate* (manipulate) her breasts in a circular motion. The purpose of this examination is to detect any developing lumps that may be lying below the surface. If a lump is discovered, a biopsy is usually performed to ascertain whether it is benign or malignant.

A biopsy is the removal of a small piece of tissue from the lump (or growth) for testing. If the lump is *benign,* surgery is the usual course of treatment. (A benign tumor is a noncancerous growth that does not invade and destroy normal tissue.) Should the lump prove to be malignant, however, the treatment may be more extensive. Frequently it necessitates complete removal of the carcinoma along with the entire breast. This surgical procedure is known as a *mastectomy*. The purpose of a mastectomy is to prevent the *metastasis,* or spread of the cancer, to nearby breast tissue.

Mastectomies are often followed by chemotherapy and/or radiation. Chemotherapy includes the use of anti-cancer drugs injected into the bloodstream. This procedure destroys any cancer cells that might have migrated into the lymph or bloodstream from the original carcinoma.

233

Uterine cancer is detected by the *Papanicolaou (PAP) smear*. Again, the usual course of treatment involves surgery. Surgical removal of the uterus is called a *hysterectomy*. Here, too, the operation may be followed by chemotherapy.

The chemotherapeutic drugs are the antimetabolites, cytotoxic or alkylating agents, antibiotics, and hormones.

An antimetabolite imitates the action of a hormone or nutrient needed by malignant cells for correct cell metabolism. Thus, the malignant cell erroneously ingests the antimetabolite and eventually dies.

Some examples of antimetabolites used to treat breast cancer are methotrexate, 5-fluorouracil, and vinblastine sulfate (Velban). Cytotoxic drugs are toxic (poisonous) to the cancer cells.

*Dysmenorrhea* is a term used for painful or difficult menstruation. Some different types of dysmenorrhea include:

- *Ovarian dysmenorrhea* — a form of dysmenorrhea due to disease of the ovaries

- *Spasmodic dysmenorrhea* — caused by severe and sudden uterine contractions

- *Uterine dysmenorrhea* — dysmenorrhea resulting from a uterine disease

Dysmenorrhea is generally characterized by abdominal pain, backache, headache, and occasionally nausea and vomiting.

*Endocervicitis* is an inflammation of the mucous membrane lining the cervix. Its key symptom is a white, muco-purulent discharge from the vagina, known as *leukorrhea*. Treatment consists of administration of antibiotics to combat the bacteria-causing cervicitis.

Fibroid tumor, or *fibroma*, is a benign tumor comprised mainly of fibrous connective tissue found in the uterus. Symptoms include backache and abnormal uterine bleeding. The general course of treatment is a *fibroidectomy*, or surgical removal of the uterine fibroma.

*Retroversion* is a disorder of the uterus. The nongravid uterus normally tilts forward about 90 degrees in proportion to the vagina. Sometimes, however, the uterus may be abnormally tilted backward, a condition called *retroversion of the uterus*. Common symptoms, if any, may include backache, constipation, or dysmenorrhea.

*Salpingitis* is inflammation of the fallopian tubes. It is accompanied by lower abdominal tenderness and pain. There are several different types of salpingitis:

- *Gonococcic salpingitis* is an infection of the fallopian tubes, caused by the **gonococcus bacterium**.

- *Purulent salpingitis* produces a purulent (pus) discharge rather than mucus or serum.

- *Tuberculosis salpingitis* is brought about by infiltration of the uterine membrane lining and wall by tuberculous nodules.

Treatment of these and related inflammatory conditions generally necessitates the use of antibiotic chemicals.

*Sterility* is the inability to reproduce: it may occur in either sex.

# MALE REPRODUCTIVE SYSTEM

*Epididymitis* is a painful swelling in the groin and scrotum due to infection of the epididymis.

*Orchitis* is inflammation of the testis. It may be a complication of mumps, influenza, or other infection. A symptom is the swelling of the scrotum, accompanied by elevated temperature and pain.

Both epididymitis and orchitis are treated by the administration of antibiotics.

*Prostatitis* is an inflammation of the prostate gland. By its pressure on the bladder, the prostate gland causes frequent, painful urination. If pressure on the urethra is severe, urinary retention may result. A *prostatectomy* is the surgical removal of all or part of the prostate gland.

## Further Study and Discussion

- A woman of forty-seven years of age asks your advice concerning depression which she has experienced for several months. She has also had difficulty sleeping due to excessive perspiration during the night. How would you advise her?

- What is the Papanicolaou test? What is its value?

## Assignment

Match each term in column I with its description in column II.

| Column I | Column II |
|---|---|
| 1. carcinoma | a. surgical removal of uterus |
| 2. endocervicitis | b. inflammation of fallopian tubes |
| 3. epididymitis | c. painful swelling in groin and scrotum |
| 4. hysterectomy | d. backward displacement of uterus |
| 5. mastectomy | e. may cause urine to be retained |
| 6. orchitis | f. surgical removal of breast |
| 7. prostatitis | g. inability to reproduce |
| 8. retroversion | h. inflammation in the lining of cervix |
| 9. salpingitis | i. cancer in breasts or uterus |
| 10. sterility | j. complication of disease or inflammation |
| | k. painful menstruation |

# SELF-EVALUATION

## Section 8 HUMAN REPRODUCTION

A. Match each term in column I with its correct description in column II.

| Column I | Column II |
|---|---|
| ___ 1. DNA | a. specialized sex cell of either sex |
| ___ 2. fertilization | b. male sex cell |
| ___ 3. gamete | c. conception |
| ___ 4. gonadotrophic hormone | d. determine hereditary characteristics |
| ___ 5. graafian follicle | e. reduction division |
| ___ 6. meiosis | f. microscopic sac |
| ___ 7. ovum | g. female sex cell |
| ___ 8. progesterone | h. stimulates testes to action |
| ___ 9. sperm | i. a fertilized cell |
| ___ 10. zygote | j. prevents menstruation during pregnancy |
| | k. study of genetics |

B. Complete the following statements.

1. The tubes which receive the ova and allow them to pass into the uterus are the _____ .

2. The cavity below the abdominal cavity is the _____ .

3. The substance called _____ carries the inherited characteristics in the chromosomes.

4. The unborn baby formed after the zygote divides is called a(an) _____ .

5. The union of the ovum and sperm cell is called _____ .

# Section 9
# Regulators of Body Functions

# Unit 37
# INTRODUCTION TO THE ENDOCRINE SYSTEM

## OBJECTIVES

- List the glands which make up the endocrine system
- Locate the endocrine glands in the body
- Describe how each endocrine gland affects body activities

A gland is any organ that produces a secretion. *Endocrine glands* are organized groups of tissues which use materials from the blood or lymph to make new compounds called *hormones*. Endocrine glands are also called ductless glands and glands of internal secretion; the hormones are secreted directly into the bloodstream as the blood circulates through the gland. The secretions are then transported to all areas of the body where they have a special influence on cells, tissues, and organs.

One of the endocrine glands, the pancreas, has two major functions. The pancreas acts as a digestive gland in the production of *pancreatic fluid*, which passes through ducts to the digestive tract. Special groups of cells, known as islets of Langerhans, secrete the hormone *insulin* which is discharged directly into the bloodstream.

There are six important endocrine glands, or groups of glands, in the body:

- pituitary gland, at the base of the brain
- thyroid gland, in the neck
- parathyroid glands, near the thyroid gland
- pancreas, behind the stomach
- two adrenal glands, one over each kidney
- gonads, or sex glands: ovaries in the female lower abdomen, testes in the male scrotum.

Figure 37-1 shows the locations of the endocrine glands in the body. Each has specific functions to perform. Any disturbance in the functioning of these glands may cause changes in the appearance or functioning of the body. Sometimes both conditions arise.

① PINEAL BODY
② PARATHYROID GLANDS
③ ADRENAL GLANDS
④ ISLETS OF LANGERHANS (pancreas)
⑤ OVARIES (female gonads)
⑥ TESTES (male gonads)
⑦ THYROID GLAND
⑧ PITUITARY GLAND
⑨ THYMUS GLAND

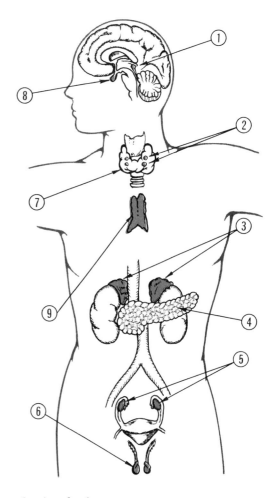

Figure 37-1 Location of the endocrine glands

## Further Study and Discussion

- Discuss location, appearance, and inter-related activities of the endocrine glands.
- Determine from outside reading what secretions of the endocrine glands can be manufactured by artificial means (synthesis).

- A boy of four has shown no growth increase since the age of two years. He seems to have a low mentality. He usually holds his mouth open, and his tongue is large. His hair is dry and coarse. Which gland possibly may be responsible for this condition?
- Discuss other hormonal secretions and glandular structures: gastrin, secretin, placental, pineal, and thymus.

## Assignment

1. Define "gland."

2. How do endocrine glands differ from other types of glands?

3. Give the general name of a secretion from an endocrine gland.

4. Name two secretions released from the pancreas.

5. Name the six important endocrine glands of the body.

6. What gland contains the islets of Langerhans?

# KEY WORDS

adrenocorticotrophic hormone
anterior lobe
follicle-stimulating hormone
hypophyseal gland
lactogenic hormone
luteinizing hormone
oxytocin
posterior lobe
prolactin
sella turcica
somatotropin
thyrotropin
vasopressin

# Unit 38

# THE PITUITARY GLAND

## OBJECTIVES

- Locate the pituitary gland
- Describe the functions of the pituitary gland
- List the principal secretions of the pituitary gland
- Define the Key Words in this unit of study

The *pituitary* gland is located at the base of the brain within the *sella turcica,* a small bony depression in the sphenoid bone of the skull. It is called the master gland because it secretes several hormones into the bloodstream which affect other glands. These glands, together with the pituitary gland secretions, help maintain proper body functioning.

The pituitary gland is responsible for the growth of the long bones; thus, it controls the height of the individual. Circus giants are often the result of overgrowth of the long bones, caused by oversecretion by the pituitary gland. The organs of reproduction are influenced by it as pituitary secretions are essential to pregnancy and lactation. The pituitary is responsible for maintaining the water balance of the body, and affects the use of starches and sugars by the body. It secretes ACTH, one of the hormones used in the treatment of arthritis. The pituitary gland is also known as the *hypophyseal* gland. It is the size of a pea and consists of the *anterior* and *posterior* lobes, figure 38-1.

It has been established that six hormones are discharged from the anterior lobe and two from the posterior. Two others appear to be discharged from an intermediate area between the two lobes. The six established hormones of the anterior lobe and the two of the posterior lobe are listed with their known functions in table 38-1.

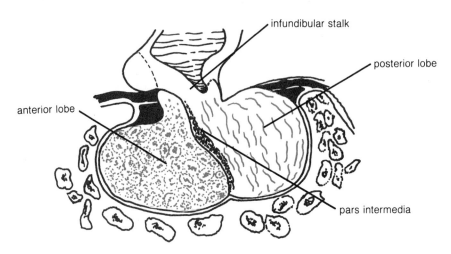

Figure 38-1 The pituitary gland

Table 38-1 Pituitary Hormones and Their Known Functions

| PITUITARY HORMONE | KNOWN FUNCTION |
|---|---|
| **Anterior Lobe** | |
| TSH — Thyroid-Stimulating Hormone (Thyrotropin) | Stimulates the growth and the secretion of the thyroid gland |
| ACTH — Adrenocorticotrophic Hormone | Stimulates the growth and the secretion of the adrenal cortex |
| FSH — Follicle-Stimulating Hormone | Stimulates growth of new graafian (ovarian) follicle and secretion of estrogen by follicle cells in the female and the production of sperm in the male |
| LH — Luteinizing Hormone (female) | Stimulates ovulation and formation of the corpus luteum |
| ICSH — Interstitial Cell-Stimulating Hormone (male) | Stimulates testosterone secretion |
| LTH — Lactogenic Hormone (Prolactin or luteotropin) | Stimulates secretion of milk and influences maternal behavior |
| GH — Growth Hormone (Somatotropin, STH) | Accelerates body growth |
| **Posterior Lobe** | |
| VASOPRESSIN (Antidiuretic Hormone, ADH) | Maintains water balance by reducing urinary output. It acts on kidney tubules to reabsorb water into the blood more quickly |
| OXYTOCIN | Promotes milk ejection and causes contraction of the smooth muscles of the uterus |

## Further Study and Discussion

- Obtain a sheep's head from the butcher or slaughterhouse. Have the lower jaw and tongue removed. Remove the floor of the skull carefully. The pituitary gland will be found in a bony encasement called the sella turcica.
- If a fresh or preserved specimen is not available, use a plastic model of the brain or a large wall chart to locate the pituitary gland.

## Assignment

A. Briefly answer the following questions.

1. Why is the pituitary gland called the master gland?

2. Describe the principal functions of the pituitary or hypophyseal gland.

3. Why may a hormone be thought of as a chemical messenger?

4. What is the name of the small depression in the sphenoid bone within which the pituitary gland is located?

5. Name the two lobes of the pituitary gland.

6. How does the pituitary gland control height in an individual?

7. Give the functions of the two hormones secreted from the posterior lobe of the pituitary gland.

8. What is the approximate size of the pituitary gland?

B. Match each of the hormones listed in column I with its function in column II. More than one function may be selected.

| Column I | Column II |
|---|---|
| _____ 1. thyroid-stimulating hormone | a. stimulates secretion of estrogen in the female and sperm production in the male |
| _____ 2. luteinizing hormone | b. stimulates testosterone secretion |
| _____ 3. follicle-stimulating hormone | c. stimulates growth and secretion of the adrenal cortex |
| _____ 4. lactogenic hormone | d. maintains water balance through kidney reabsorption |
| _____ 5. interstitial cell-stimulating hormone | e. stimulates secretion of milk in the mother |
| _____ 6. adrenocorticotrophic hormone | f. accelerates body growth |
| _____ 7. growth hormone | g. stimulates both the growth and secretion of the thyroid gland |
| _____ 8. somatotropin | |
| _____ 9. thyrotropin | h. causes contraction of smooth muscle in the uterus |
| _____ 10. prolactin | |

## KEY WORDS

calciferol (vitamin D)
calcitonin
hypercalcemia
isthmus of thyroid gland
osteoblast
osteoclast
parathormone
parathyroid gland
thymus
thyroglobulin
thyroid gland, intermediate lobe
TSH (thyroid-stimulating hormone)
thyroxine
triiodothyronine

# Unit 39
# THE THYROID AND PARATHYROID GLANDS

## OBJECTIVES

- Locate the thyroid, parathyroid, and thymus glands
- Describe the important functions of the thyroid gland
- Describe the functions of the parathyroid and thymus glands
- Define the Key Words related to this unit

The thyroid and parathyroid glands are located in the neck, close to the cricoid cartilage ("Adam's apple"). The thyroid regulates body metabolism. The parathyroid maintains the calcium-phosphorus balance.

The thymus gland is located anterior to and above the heart. It is involved with the development of an immune response in the newborn. The gland grows until puberty at which time the original lymphoid tissue is replaced with adipose and connective tissue.

## THYROID GLAND

The *thyroid gland* is a butterfly-shaped mass of tissue located in the anterior part of the neck, figure 39-1. It lies on either side of the larynx, over the trachea. Its general shape is that of the letter "H." It is about two inches long, with two lobes joined by strands of thyroid tissue called the *isthmus.* Coming from the isthmus is a finger-like lobe of tissue known as the *intermediate lobe.* This intermediate lobe projects upward toward the floor of the mouth, as far up as the hyoid bone. The thyroid gland has a rich blood supply. In fact, it has been estimated that about 4 to 5 liters (some 8 1/2 to 10 1/2 pints) of blood pass through the gland every hour.

The thyroid gland secretes three hormones: *thyroxine, triiodothyronine,* and *calcitonin.* The first two are iodine-bearing derivatives of the amino acid, tyrosine. Triiodothyronine is 5 to 10 times more active than thyroxine, but its activity is less prolonged. However, the two have the same effect. Both hormones are produced in the follicle cells of the thyroid gland. These cells are stimulated to secretory activity by a hormone from the anterior lobe of the pituitary gland. This hormone, *TSH (thyroid-stimulating hormone),* controls the production

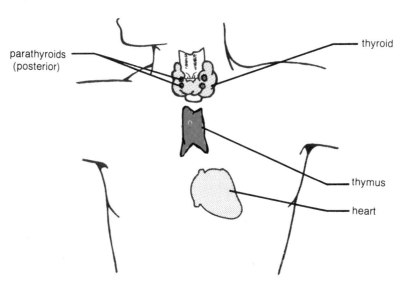

Figure 39-1 Location of the thyroid, parathyroid, and thymus glands

of the thyroid hormones. It does so by ingesting iodine* through food or water, which is eventually absorbed by the intestines into the bloodstream. The iodides are circulated to the thyroid gland, where they are "trapped." Here the iodides combine with the amino acid tyrosine to form the thyroglobulin molecule. Prior to being secreted into the bloodstream, thyroglobulin is chemically converted into triiodothyronine ($T_3$), and finally into thyroxine ($T_4$).

Under normal circumstances, the presence of these two hormones ($T_3$ and $T_4$) in the bloodstream serves to regulate the system. On the other hand, an excess would suppress TSH secretion. Consequently, the thyroid gland secretes less of these two hormones. When the concentration of thyroid hormones is lowered in the bloodstream, the pituitary gland secretes more TSH. This, in turn, stimulates thyroid gland activity. (The consequences of hyposecretion and hypersecretion of the thyroid hormones is discussed in Unit 42, *Representative Disorders of the Endocrine System*).

Thyroxine controls the rate of metabolism, heat production, and oxidation of all cells, with the possible exception of the brain and spleen cells. We have already noted the thyroid's role in regulating physical growth, mental development, sexual maturity, and the distribution and exchange of water and salts within the body. It can speed up or slow down the activities of the body as needed. In the liver, the two thyroid hormones affect the conversion of glycogen from sources other than sugar. It also helps to change glycogen into glucose, raising the glucose level of the blood.

## Calcitonin

*Calcitonin* is another hormone produced and secreted by the thyroid gland. It controls

---

*In its pure elemental form, iodine is a violent cellular poison. It exists, however, within the human body in minute amounts as iodides. Half of the body's total iodine supply (about 50 milligrams) is found in the thyroid as *thyroglobulin* (a combination of tyrosine and iodine).

the calcium ion concentration in the body by maintaining a proper calcium level in the bloodstream.

Calcium is an essential body mineral. Approximately 99% of the calcium in the body is stored in the bones. The rest is located in the blood and tissue fluids. Calcium is necessary for blood clotting, holding cells together, and neuromuscular functions. The constant level of calcium in the blood and tissues is maintained by the action of calcitonin and parathormone (produced by the parathyroid gland).

Calcitonin lowers the calcium concentration in the blood and body fluids by acting on bone cells known as *osteoblasts* (specialized bone cells that form the bony material of the skeleton). When blood calcium levels are higher than normal, calcitonin secretion is increased. This mechanism prompts calcium removal from the blood, depositing it within the bones.

Insufficient deposition of calcium in children's bones will lead to rickets. The bone disorder is due to the lack of a hormone called *calciferol,* or *vitamin D.* Calciferol stimulates the intestine to absorb calcium. Proper secretion of calcitonin into the bloodstream prevents a harmful rise in the blood calcium level, or *hypercalcemia.*

## PARATHYROID GLANDS

The *parathyroid* glands, usually four in number, are tiny glands the size of grains of rice. These are attached to the posterior surface of the thyroid gland, and secrete the hormone, *parathormone.* Parathormone, like calcitonin, also controls the concentration of calcium in the bloodstream. When the blood calcium level is lower than normal, parathormone secretion is increased.

Parathormone stimulates an increase in the number and size of specialized bone cells referred to as *osteoclasts.* Osteoclasts quickly invade hard bone tissue, digesting large amounts of the bony material containing calcium. As this process continues, calcium leaves the bone and is released into the bloodstream, increasing the calcium blood level.

Bone calcium is bonded to phosphorus in a compound called calcium phosphate ($CaPO_4$). So when calcium is released into the bloodstream, phosphorus is released along with it. Parathormone stimulates the kidneys to excrete any excess phosphorus from the blood; at the same time, it inhibits calcium excretion from the kidneys. Consequently, the concentration of blood calcium rises.

Thus parathormone and calcitonin have opposite, or antagonistic effects to one another (see figure 39-2 for a summary of their actions). Parathormone, however, acts much more slowly than calcitonin. It may be hours before the effects of parathormone become apparent. In this manner, the secretion of parathormone and calcitonin serve as complementary processes controlling the level of calcium in the bloodstream.

## THYMUS GLAND

The *thymus* gland is located under the breastbone, or sternum. Fairly large during childhood, it begins to disappear at puberty. Little is known about the gland, but it is believed that some kind of thymus hormone, *thymosin,* is secreted from this organ during infancy. This stimulates the lymphoid cells which are responsible for the production of antibodies against certain diseases.

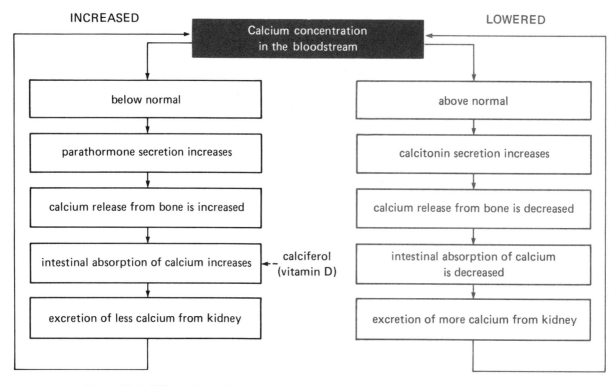

Figure 39-2 Effects of parathormone and calcitonin on the level of calcium in the blood

## Further Study and Discussion

- Locate the thyroid, parathyroids, and thymus glands using classroom models or wall charts.
- Relate the function of the thyroid gland to the process of metabolism.

## Assignment

1. Locate and describe the thyroid gland.

UNIT 39 THE THYROID AND PARATHYROID GLANDS  249

2. Locate and describe the parathyroid glands.

3. Describe the functions of the thyroid, parathyroid, and thymus glands.

4. Label the diagram below.

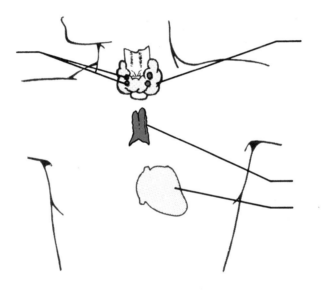

# Unit 40
# THE ADRENAL GLANDS AND GONADS

## KEY WORDS

androgens
epinephrine
glucocortocoids
mineralcorticoids
progesterone
testosterone

## OBJECTIVES

- Locate the adrenal glands and gonads
- Describe functions of the adrenals and gonads
- Name the secretions of the adrenals and gonads
- Define the Key Words related to this unit of study

One of the two *adrenal* glands is located on top of each kidney, figure 40-1. Each gland has two parts, the *cortex* and the *medulla*. Adrenocorticotrophic hormone (ACTH) from the pituitary glands stimulates the activity of the cortex of the adrenal gland. The hormones secreted by the adrenal cortex are known as *corticoids*. The corticoids are very effective as anti-inflammatory drugs.

The cortex secretes three groups of corticoids, each of which is of great importance:

- *Mineralcorticoids* (M-Cs) affect the kidney tubules by speeding up the reabsorption of sodium into the blood circulation and increasing the excretion of potassium from the blood. They also speed up the reabsorption of water by the kidneys. Aldosterone (M-C) is used in the treatment of Addison's Disease to replace deficient secretion of mineralcorticoids.

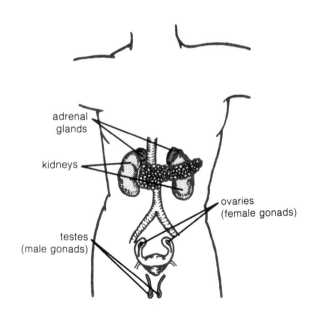

Figure 40-1 Location of adrenals and gonads

- *Glucocorticoids* (G-Cs) increase the amount of glucose in the blood. This is presumably done by (1) conversion of the protein brought to the liver into glycogen, followed by (2) breakdown of the glycogen into glucose. These glucocorticoids also help the body resist the aggravations caused by various everyday stresses. Both cortisone and hydrocortisone are G-Cs.

- *Androgens* are male sex hormones which, together with similar hormones from the gonads, bring about masculine characteristics. They also promote protein anabolism and body growth.

The medulla of the adrenal gland secretes *epinephrine* and *norepinephrine.* Epinephrine (generic name), or Adrenalin (trade), is a powerful cardiac stimulant. It functions by bringing about a release of more glucose from stored glycogen for muscle activity; and increasing the force and rate of the heartbeat. This chemical activity increases cardiac output and venous return, and raises the systolic blood pressure.

The gonads, or sex glands, include the ovaries in the female and the testes in the male. The ovaries are located in the pelvic cavity, one on either side of the uterus. The testes are located in the scrotum.

The secretions of the ovaries, *estrogen* and *progesterone,* are necessary for ovulation and the characteristic female appearance. The secretion of the testes, *testosterone,* is essential to the development of the male sex characteristics. The gonads are responsible for fertility and reproduction in both sexes. Review Unit 35 for further details about the reproductive organs.

## Further Study and Discussion

- There are many instances on record of individuals performing superhuman feats during emergencies. Perhaps you have had such an experience yourself. What is the explanation for such strength at these times?

- Secure a lamb kidney with the adrenal gland attached. Make a long incision through the adrenal gland. Notice that it has an outer area and an inner area. Each part contributes its own secretion. Compare the lamb's adrenal gland with a picture or model of the human gland, noticing shape, size, and color.

- If frogs are available to you for laboratory work, inject about 9.2 milliliters of adrenalin chloride (1:100 000 solution) into the ventricle of the heart of an anesthetized frog. Count the heartbeat before the injection and after the injection. What did the adrenalin do to the heart? What was the quality of the heartbeat before and after injection?

- Describe the use of adrenalin for hemorrhage and asthma control.

## Assignment

1. Describe the location of the adrenal glands.

2. Summarize the functions and secretions of the adrenals and the gonads, using the following chart.

| GLAND | SECRETION | FUNCTION |
|---|---|---|
| Adrenals | Cortex | |
| | Medulla | |
| Gonads | Ovaries | |
| | Testes | |

# Unit 41
# THE PANCREAS

OBJECTIVES

- Describe the endocrine functions of the pancreas
- Explain the body's need for insulin

The *pancreas* is an organ located behind the stomach. The glandular cells of the pancreas regulate the production of pancreatic juice, a digestive juice. The islet cells secrete the hormone *insulin*. Therefore the pancreas is a gland of both external and internal secretion.

The islet cells (mostly beta cells) are distributed throughout the pancreas. These cells were named the islets of Langerhans

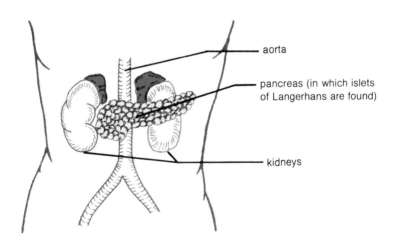

Figure 41-1 Location of islets of Langerhans

after the doctor who discovered them. Beta cells produce insulin which: (1) promotes the utilization of glucose in the cells, necessary for maintenance of normal levels of blood glucose, (2) promotes fatty acid transport and fat deposition into cells, (3) promotes amino acid transport into cells, and (4) facilitates protein synthesis. Lack of insulin secretion by the island (islet) cells causes diabetes mellitus.

## Further Study and Discussion

- If laboratory facilities are available, obtain a sheep or beef pancreas. Prepare a thin section for use with the microscope.
    a. Focus the slide under low power to show the arrangement of the islets. Discuss what you see.
    b. Focus the slide under high power to observe the arrangement of the cells in the islets and the pancreas in general.
- Obtain several specimens of urine to test for the presence of sugar. Pour 5 milliliters (cc) of Benedict's solution into a test tube. Using a medicine dropper, add 10 drops of urine to the solution. Shake well to mix thoroughly. Boil the solution in a test tube in water bath for 3-5 minutes. Note the color.

| COLOR | INDICATION |
| --- | --- |
| Blue | Sugar free |
| Green | + present, trace |
| Greenish yellow | 1+ |
| Yellow | 2+ |
| Brown or red | 3+ to 4+ |

Your instructor may have new tablet testing devices for the presence of sugar. Try these simplified methods as well as the Benedict's solution method.

- Research and discuss whether insulin may be considered a cure for diabetes.

## Assignment

1. What is meant by the statement that the pancreas is a gland of external as well as internal secretion?

2. Where is the pancreas located?

3. Where in the pancreas is insulin produced?

4. What are the functions of insulin?

5. What condition is caused by a lack of insulin in the bloodstream?

# Unit 42
# REPRESENTATIVE DISORDERS OF THE ENDOCRINE SYSTEM

## KEY WORDS

acidosis
acromegaly
Addison's disease
cretin
cretinism
Cushing's syndrome
diabetes mellitus
electrolyte
exophthalmos
gigantism
glycosuria
goiter
hyperthyroidism
ketone body
myxedema
pituitary dwarfism
tetany

## OBJECTIVES

- List causative factors of endocrine gland disorders
- Recognize certain endocrine gland disorders which interfere with body functions
- Relate treatment to some of the more common types of gland disorders
- Define the Key Words related to this unit of study

Endocrine gland disturbances may be caused by several factors, such as disease of the gland itself, infections in other parts of the body, and dietary deficiencies. Most disturbances result from (1) hyperactivity of the glands, causing oversecretion of hormones, or (2) hypoactivity of the gland, resulting in undersecretion of hormones.

## THYROID DISORDERS

Since the thyroid gland controls metabolic activity, any disorder will affect other structures besides the gland itself. Signs and symptoms of the disorders most frequently seen are discussed in this unit.

### Hyperthyroidism

*Hyperthyroidism* is due to the overactivity of the thyroid gland. Thus too much thyroxin is secreted (hypersecretion), leading to enlargement of the gland. An increase in the basal metabolic rate (BMR) also occurs. In hyperthyroidism, an individual's BMR may be 50 to 75% higher than normal. People with hyperthyroidism consume large quantities of food, but nevertheless suffer a loss of body fat and weight. They may suffer from increased blood pressure and heartbeat, hand tremors, perspiration, and irritability. In addition, the liver releases excess glucose into the bloodstream, increasing the blood sugar level and causing a mild case of *glycosuria.*

UNIT 42   REPRESENTATIVE DISORDERS OF THE ENDOCRINE SYSTEM   257

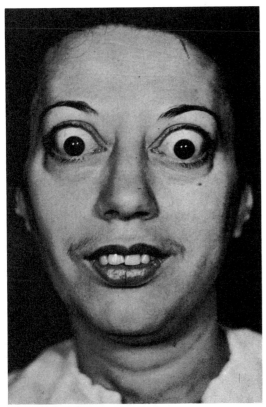

Figure 42-1 Hyperthyroidism (Reproduced with permission from S.L. Robbins, and R.S. Cotran, *Pathological Basis of Disease.* Philadelphia: W.B. Saunders.)

Determination of the basal metabolism rate (BMR) is an aid in diagnosing thyroid function. The test is performed by measuring the amount of oxygen a person uses at rest; from this the basal body metabolism rate can be calculated. Another diagnostic test is the protein-bound iodine test (P.B.I.). This blood test determines the concentration of thyroxin in the bloodstream. Other thyroxin level tests are the $T_3$ and $T_4$. These tests are more accurate than the P.B.I. but are more difficult to perform.

The *radioactive iodine uptake* test measures the activity of the thyroid gland. Dilute radioactive iodine is given orally. The amount which accumulates in the thyroid gland is calculated by use of a Geiger counter. Simple goiter is an enlargement of the thyroid gland due to a deficiency of iodine in the diet.

The most pronounced symptoms of hyperthyroidism include enlargement of the thyroid gland (goiter), bulging of the eyeballs (exophthalmos), dilation of the pupils, and wide-opened eyelids.

The immediate cause of exophthalmos is not completely known. It is not directly caused by the hyperthyroidism, because removal of the thyroid does not always cause the eyeballs to return to their normal state. Treatment of hyperthyroidism includes total or partial removal of the thyroid, and administration of drugs like propylthiouracil and methylthiouracil to reduce the thyroxin secretion.

Hypothyroidism

*Hypothyroidism* is a condition in which the thyroid gland does not secrete sufficient thyroxin (hyposecretion). This is manifested by a low metabolic rate and decelerated body processes. Depending upon the time hypothyroidism strikes its victims, two different sets of disorders may occur, cretinism or myxedema.

**Cretinism.** This disorder develops in early infancy or childhood. It is characterized by a lack of mental and physical growth, resulting in mental retardation and malformation (dwarfism or *cretinism*). The sexual development and physical growth of cretins do not proceed beyond that of 7- or 8-year-old children.

In treating cretinism, thyroid hormones (or sodium liothyronine) may restore a degree of normal development if administered in time. In most cases, however, normal development cannot be completely restored once the affliction has set in.

**Myxedema.** This condition is similar to cretinism, although it develops during later childhood or in adult life. *Juvenile myxedema* commonly results in a short, squat physique. The young person suffers from dry skin, an enlarged head, a short and heavy neck, and a dull and vacant facial expression.

*Adult myxedema* produces edema, lethargy, obesity, decreased heartbeat and lowered intelligence. The hair and skin grow coarse. Victims tend to feel cold due to a greatly diminished metabolic rate.

The treatment for thyroid deficiency usually consists of administering oral thyroid extract so as to restore a normal metabolism.

## PARATHYROID DISORDERS

The parathyroid glands regulate the use of calcium and phosphorus. Both of these minerals are involved in many of the body systems.

Hyperfunctioning of the parathyroid glands may cause an increase in the amount of blood calcium, thereby increasing the tendency for the calcium to crystallize in the kidneys as *kidney stones*. Excess amounts of calcium and phosphorus are withdrawn from the bones; this may lead to eventual deformity. So much calcium can be removed from the bones that they become honeycombed with cavities. Afflicted bones become so fragile that even walking can cause fractures.

Hypofunctioning of the parathyroid glands leads to a condition known as *tetany*. In this case, severely diminished calcium levels affect the normal function of nerves. Convulsive twitchings develop, and the afflicted person dies of spasms in the respiratory muscles. Treatment consists of administering vitamin D, calcium, and parathormone to restore a normal calcium balance.

## PITUITARY DISORDERS

Disturbances of the pituitary gland may produce a number of body changes. This gland is chiefly involved in the growth function. However, as the master gland, the pituitary indirectly influences other activities.

### Gigantism

Hyperfunctioning of the pituitary gland (often due to a pituitary tumor) causes hypersecretion of the pituitary growth hormone. When this occurs during preadolescence it causes *gigantism,* an overgrowth of the long bones leading to excessive tallness. The most famous giant was Robert Wadlow (1919-1940) of Alton, Illinois. Wadlow grew to 8'-10 3/4" in height and weighed 495 lbs. (He wore a size 37 shoe!)

If hypersecretion of the growth hormone occurs during adulthood, *acromegaly* results. This is an overdevelopment of the bones of the face, hands and feet. In adults whose long bones have already matured, the growth hormone attacks the cartilaginous regions and the bony joints. Thus the chin protrudes, and the lips, nose, and extremities enlarge disproportionately. Lethargy and severe headaches frequently set in as well.

Treatment of acromegaly and gigantism is difficult due to the inaccessibility of the pituitary gland. If a pituitary tumor is responsible for the hypersecretion of growth hormones, surgery or bombardment with X rays may offer some relief.

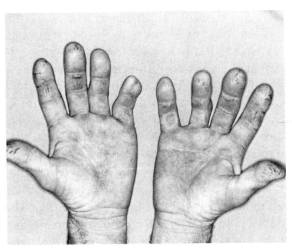

Figure 42-2 Effects of acromegaly on fingers and hands (Armed Forces Institute of Pathology, negative 72-14615)

Hypofunctioning of the pituitary gland during childhood leads to *pituitary dwarfism*. Growth of the long bones is abnormally decreased by an inadequate production of growth hormone. Despite the small size, however, the body of a dwarf is normally proportioned and intelligence is normal. Unfortunately, the physique remains juvenile and sexually immature. Treatment involves early diagnosis and injections of human growth hormone. The treatment period is 5 years or more.

Other disorders caused by pituitary hypofunctioning include diabetes insipidus and menstrual problems.

## ADRENAL DISORDERS

Overactivity of the adrenal gland may result in virilism and Cushing's syndrome. Virilism is the development of male secondary characteristics in a woman (facial hair, broad shoulders, small breasts).

*Cushing's syndrome* results from the hypersecretion of the glucocorticoid hormone from the adrenal cortex. This hypersecretion may be caused by an adrenal cortical tumor. (Oddly enough, more women than men tend to develop this endocrine disorder.) Symptoms include high blood pressure, muscular weakness, obesity, poor healing of skin lesions, and a tendency to bruise easily. The most noticeable characteristics are a rounded "moon" face and a "buffalo hump" that develops from the redistribution of body fat. Therapy consists of surgical removal of the adrenal cortical tumor.

Hypofunctioning of the adrenal cortex can also lead to *Addison's disease*. Persons with Addison's disease exhibit the following symptoms:

- Excessive pigmentation, prompting the characteristic "bronzing" of the skin
- Decreased levels of blood glucose, causing carbohydrate imbalance
- A severe drop in blood pressure, leading to kidney malfunction
- Pronounced muscular weakness and fatigue
- Gastrointestinal malfunction, resulting in diarrhea, weight loss and vomiting
- Retention of water in the body tissues
- A severe drop of sodium in the blood and tissue fluids, causing a serious imbalance of *electrolytes*

Patients are treated by administration of regular doses of cortisone and a controlled intake of sodium.

## GONAD DISORDERS

Disturbances in the ovaries may consist of cysts and tumors, abnormal menstruation, and menopausal changes. Turner's syndrome may occur in either the male or female; this is a chromosomal disorder.

## PANCREATIC DISORDERS

*Diabetes mellitus* is a condition caused by decreased secretion of insulin from the islet cells of the pancreas. As a result, carbohydrate metabolism is disturbed; this has an adverse effect on protein and fat metabolism.

Insulin deficiency causes glucose to accumulate in the bloodstream, rather than be transported to the cells and converted into energy. Eventually the excess becomes too much for the kidneys to reabsorb, and the excess glucose is excreted in the urine. Excretion of excess glucose requires an accompanying excretion of large amounts of water. This occurs to insure that the sugar concentration does not rise too high. Diabetics are constantly thirsty because the lost water must be replaced.

Since insufficient glucose is available for cellular oxidation in such cases, the body starts to burn up protein and fats. The diabetic is constantly hungry and usually eats voraciously, but loses weight nonetheless.

When fats are utilized as a fuel source, they are rapidly but incompletely oxidized. One product of this abnormal rate of fat oxidation is *ketone bodies*. Ketone bodies are highly toxic; the type most commonly formed is acetoacetic acid. These keto acids accumulate in the blood, promoting the development of *acidosis,* giving the breath and urine an odor of "sweet" acetone. If acidosis is severe, *diabetic coma* and death may result. Prolonged diabetes leads to atherosclerosis, heart disease, and kidney damage. Therapy consists of daily insulin injections, or (in some cases) daily antidiabetic tablets and a restricted carbohydrate and salt diet. Most diabetics live active, normal lives when properly treated.

---

## Further Study and Discussion

- Discuss how a disturbance in the pituitary gland can affect the functioning of other endocrine glands.
- Discuss some of the symptoms of diabetes mellitus.

## Assignment

1. What is the meaning of each of the following terms?

    a. hyperfunctioning

    b. hypofunctioning

2. What symptoms indicate hyperfunctioning of the thyroid gland?

3. What conditions may result from hypofunctioning of the pituitary gland?

4. Complete the following chart.

| GLAND | HORMONE | NORMAL FUNCTION | DISORDERS |
|---|---|---|---|
| Pituitary | | | |
| Thyroid | | | |
| Parathyroids | | | |
| Thymus | | | |
| Adrenals | | | |
| Gonads | | | |
| Pancreas | | | |

# SELF-EVALUATION

## Section 9 REGULATORS OF BODY FUNCTIONS

A. Match each term in column I with its correct description or function in column II.

| Column I | Column II |
|---|---|
| \_\_\_ 1. ACTH | a. master gland of the endocrine system |
| \_\_\_ 2. adrenals | b. any gland of internal secretion |
| \_\_\_ 3. cortisone | c. a hormone secreted by adrenals |
| \_\_\_ 4. gonad | d. regulates use of calcium |
| \_\_\_ 5. endocrine | e. the secretion of any endocrine gland |
| \_\_\_ 6. hormone | f. helps body meet emergencies |
| \_\_\_ 7. insulin | g. sex gland |
| \_\_\_ 8. parathyroid | h. regulates body metabolism |
| \_\_\_ 9. pituitary | i. one of the hormones secreted by pituitary gland |
| \_\_\_ 10. thyroid | j. a hormone which regulates carbohydrates and metabolism |
|  | k. hypofunction of endocrine glands |

B. Identify the labelled glands.

① _____
② _____
③ _____
④ _____
⑤ _____
⑥ _____
⑦ _____
⑧ _____
⑨ _____

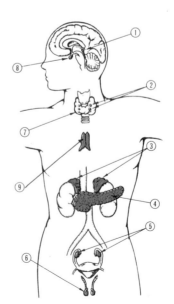

# Section 10
# Coordination of Body Functions

# Unit 43
# INTRODUCTION TO THE NERVOUS SYSTEM

## KEY WORDS

| | | |
|---|---|---|
| associative (connecting) neuron | conductivity | neuron |
| autonomic nervous system | dendrites | peripheral nervous system |
| axon | impulse | sensory neuron |
| central nervous system | irritability | |
| | motor neuron | |

## OBJECTIVES

- Describe the functions of the nervous system
- List the main parts of the nervous system
- Describe three types of neurons
- Define characteristics of the nerve cells
- Define the Key Words related to this unit of study

The study of body functions reveals that the body is made up of millions of small structures that perform a multitude of different activities; these are coordinated and integrated into one harmonious whole. The two main communication systems are the endocrine system and the nervous system. They send chemical messengers and nerve impulses to all of the structures. The endocrine system and hormonal regulation have been discussed in earlier units. Hormonal regulation is slow, while neural regulation is comparatively rapid.

The nervous system is the most highly organized system of the body, consisting of the brain, spinal cord, and nerves. The structural and functional unit, as in other systems, is the cell. The nerve cell, or *neuron*, is especially constructed to carry out its function of communication. In addition to the nucleus, cytoplasm, and cell membrane, the neuron has extensions of cytoplasm from the cell body. These extensions, or processes, are called *dendrites* and *axons*. There may be several dendrites, but only one axon. These processes, or *fibers*, as they are often called, are paths along which nerve impulses travel, figure 43-1.

All neurons possess the characteristics of being able to react when stimulated and of being able to pass the nerve impulse generated on to other neurons. These characteristics are *irritability* (the ability to react when stimulated) and *conductivity* (the ability to transmit a disturbance to distant points). The dendrites receive the impulse and transmit it to the cell body, and then to the axon where

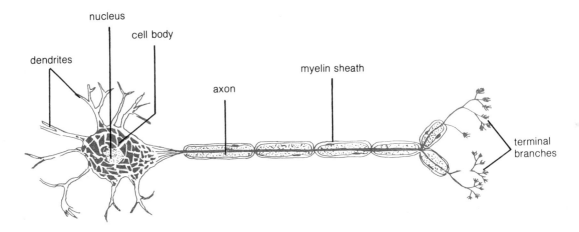

Figure 43-1 A nerve cell

it is passed on to another neuron or to a muscle or gland. There are three types of neurons:

- *Sensory neurons* which emerge from the skin or sense organs and carry messages, or impulses, toward the spinal cord and brain
- *Motor neurons* which carry messages from the brain and spinal cord to the muscles and glands
- *Connecting, or associative neurons* which carry impulses from one neuron to another

The nervous system can be divided into three divisions: the central, the peripheral, and the autonomic nervous system.

1. The *central nervous system* consists of the brain and spinal cord
2. The *peripheral nervous system* is made up of the nerves of the body consisting of twelve pairs of cranial nerves extending out from the brain and thirty-one pairs of spinal nerves extending out from the spinal cord
3. The *autonomic nervous system* is comprised of a chain of ganglia (group of neuron cell bodies) on either side of the spinal cord

Where decision is called for and action must be considered, the central and peripheral nervous systems are involved. They carry information to the brain where it is interpreted, organized, and stored. An appropriate command is sent to organs or muscles. The autonomic nervous system supplies heart muscle, smooth muscle, and secretory glands with nervous impulses as needed. It is usually involuntary in action.

## Assignment

Complete the following statements.

1. Two types of muscle tissue supplied with nerve impulses from the autonomic nervous system are _____ and _____.

2. Neurons which emerge from the skin or sense organs and carry messages toward the spinal cord are called _____ neurons.

3. Neurons which carry messages from the brain and spinal cord to the muscles and glands are called _____ neurons.

4. The ability of a neuron to react when stimulated is called _____.

5. The extension of the nerve cell body which receives the impulse is called the _____ and the part which passes the impulse on is called the _____ .

6. Glands that receive stimulation from the autonomic nervous system are called _____ glands.

# KEY WORDS

| | | |
|---|---|---|
| arachnoid | diencephalon | longitudinal fissure |
| arachnoid villi | (interbrain) | lumbar puncture |
| blood-brain barrier | dura mater | medulla |
| brainstem | ependymal cells | meninges |
| central fissure (fissure of Rolando) | exudate | occipital lobe |
| | fissure | parietal lobe |
| cerebellum | foramen of Magendie | parieto-occipital fissure |
| cerebral aqueduct (aqueduct of Sylvius) | fourth ventricle | pia mater |
| | frontal lobe | pons |
| cerebral ventricles | gyrus (plural, gyri) | sensory (somesthetic) area |
| cerebrospinal fluid | hypothalamus | |
| cerebrum | interventricular foramen | sulcus (plural, sulci) |
| choroid plexus | | temporal lobe |
| convolutions | lateral fissure (fissure of Sylvius) | thalamus |
| corpus callosum | | third ventricle |
| cortex | lateral ventricle | transverse fissure |

# Unit 44
# THE CENTRAL NERVOUS SYSTEM: BRAIN AND SPINAL CORD

## OBJECTIVES

- Identify the parts of the brain
- Describe the structure of the brain and spinal cord
- Relate functions to the various parts of the brain
- Identify the Key Words related to this unit of study

The adult human brain is a highly developed, complex, and intricate mass of soft nervous tissue. It weighs about 1400 grams (3 lbs). The brain is protected by the bony cranial cavity; further protection is afforded by three membranous coverings called *meninges*, and the cerebrospinal fluid.

The three meninges are the dura mater, arachnoid, and the pia mater, see color plate 17. The *dura mater* is the outer brain covering, which lines the skull on one side. This is a tough, dense membrane of fibrous connective tissue, containing an abundance of blood vessels. The *arachnoid (mater)* is the middle layer. It resembles a fine cobweb with fluid-filled spaces. Covering the brain surface itself is the *pia mater,* comprised of blood vessels held together by fine areolar connective tissue. The space between the arachnoid and pia mater is filled with cerebrospinal fluid, produced within the ventricles of the brain. This fluid acts both as a shock absorber and a source of nutrients for the brain.

## VENTRICLES OF THE BRAIN

The brain contains four lined cavities filled with cerebrospinal fluid. These cavities are called cerebral *ventricles.* Each ventricle is lined with ependymal cells. The ventricles lie deep within the brain. The two largest, located within the cerebral hemispheres, are known as the right and left *lateral ventricles.* Each lateral ventricle is subdivided into four parts, or horns. An *anterior horn* is located in the frontal lobe; a middle, or *body,* is found in the parietal lobe; an *inferior,* or *temporal,*

267

horn is positioned in the temporal lobe; and an *occipital,* or *posterior,* horn is situated in the occipital lobe.

The third ventricle is placed behind and below the lateral ventricles. It is connected to the two lateral ventricles via the *interventricular foramen,* or foramina of Monro.[1] The fourth ventricle is situated below the third, in front of the cerebellum, and behind the pons and the medulla oblongata. The third and fourth ventricles are interconnected via a narrow canal called the *cerebral aqueduct,* or aqueduct of Sylvius.[2] In the roof of the fourth ventricle is an opening known as the foramen of Magendie.[3] The lateral wall of the fourth ventricle contains two openings called the foramina of Luschka.[4]

Each of the four ventricles contains a rich network of blood vessels of the pia mater referred to as the *choroid plexus.* The choroid plexus is in contact with the ependymal cells lining the ventricles, which helps in the formation of cerebrospinal fluid.

Cerebrospinal Fluid
and Its Circulation

Cerebrospinal fluid is a substance that forms inside the four brain ventricles from the blood vessels of the choroid plexuses. This fluid serves as a liquid shock absorber protecting the delicate brain and spinal cord. It is formed by filtration from the intricate capillary network of the choroid plexuses. The fluid transports nutrients to, and removes metabolic waste products from, the brain cells.

Choroid plexus capillaries differ significantly in their selective permeability from capillaries in other areas of the body. A potential result is that drugs within the bloodstream may not effectively penetrate brain tissue, rendering infections (such as meningitis) difficult to cure. This phenomenon is commonly referred to as the *blood-brain barrier.*

After filling the two lateral ventricles of the cerebral hemispheres, the cerebrospinal fluid seeps into the third ventricle via the foramen of Monro. From here it flows through the aqueduct of Sylvius into the fourth ventricle. The fluid then passes through the foramen of Magendie and the two lateral foramina of Luschka, into the small, tubelike central canal of the cord and into the subarachnoid spaces. The subarachnoid spaces are thus filled with cerebrospinal fluid which bathes the brain and the spinal cord. Ultimately the cerebrospinal fluid returns to the bloodstream via the venous structures in the brain, called *arachnoid villi.*

The formation and circulation of cerebrospinal fluid is used by members of the health team to detect any defects or disease of the brain. For example, inflammation of the cranial meninges quickly spreads to the meninges of the spinal cord. This leads to an increased secretion of cerebrospinal fluid which collects in the confined bony cavity of the brain and spinal column. The accumulation of excess fluid causes headaches, reduced pulse rate, slow breathing, and partial or total unconsciousness.

Removal of cerebrospinal fluid for diagnostic purposes is accomplished with a *lumbar puncture.* The needle used to withdraw the cerebrospinal fluid is inserted between the third and fourth lumbar vertebrae. The fluid, or exudate withdrawn contains

---

1. Alexander Monro (Primus) (1697-1767), Scottish anatomist
2. Francois Sylvius (1614-1672), French anatomist
3. Francois Magendie (1783-1855), French physiologist
4. Hubert von Luschka (1820-1875), German anatomist

by-products of the inflammation and the organisms causing it. Therefore, a lumbar puncture is helpful in diagnosing such diseases as cerebral hemorrhage, increased pressure, intracranial tumors, meningitis, and syphilis. It also serves to alleviate the pressure caused by meningitis, and especially hydrocephalus. Occasionally it may be used for the introduction of antimeningitis sera or drugs.

Brain tissue is made up of gray and white matter. The outer surface of the brain is grayish, the center is white. The cortex is the highest center of the brain and is made up of so-called gray matter. The gray matter really consists of millions of nerve cell bodies and naked nerve fibers. The white matter contains millions of nerve cell fibers with myelin sheaths, which accounts for the difference in appearance.

The brain is divided into three parts: the cerebrum, cerebellum, and brainstem. The brainstem is further divided into three parts, the medulla, pons, and the midbrain, see color plate 17.

## CEREBRUM

The cerebrum is the largest and highest part of the brain. It occupies the whole upper part of the skull and weighs about two pounds. Covering the upper and lower surfaces of the cerebrum is a layer of gray matter called the *cerebral cortex*.

The cerebrum is divided into two hemispheres, right and left, by a very deep groove known as the *longitudinal fissure*. The cerebral surface is completely covered with furrows and ridges. The deeper furrows, or grooves, are referred to as *fissures,* and the shallower ones, *sulci*.

The elevated ridges between the sulci are the *gyri*, also known as *convolutions,* figure 44-1. These convolutions serve to increase the surface area of the brain, resulting in a proportionately larger amount of gray matter. The arrangement of the gyri and sulci on the brain's surface varies from one brain to another. Certain fissures, however, are constant and represent important demarcations. They help to localize specific functional areas of the cerebrum, and to divide the hemispheres into four lobes.

Each cerebral hemisphere is divided into a *frontal, parietal, occipital,* and *temporal* lobe. These lobes correspond to the cranial bones by which they are overlaid, figure 44-1.

The five major fissures dividing the cerebral hemispheres include:

- *Longitudinal fissure* — a deep groove divides the cerebrum into two hemispheres. The middle region of the two hemispheres is held together by a wide band of axonal fibers called the corpus callosum.

- *Transverse fissure* — divides the cerebrum from the cerebellum

- *Central fissure, or fissure of Rolando* — located beneath the coronal suture of the skull, dividing the frontal from the parietal lobes

- *Lateral fissure or fissure of Sylvius* — situated on the side of the cerebral hemispheres, dividing the frontal and temporal lobes

- *Parieto-occipital fissure* — the least obvious of all the fissures, serves to separate the occipital lobe from the parietal and temporal lobes. There is, however, no definite demarcation between these two lobes.

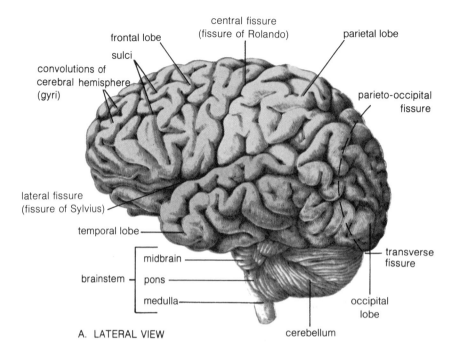
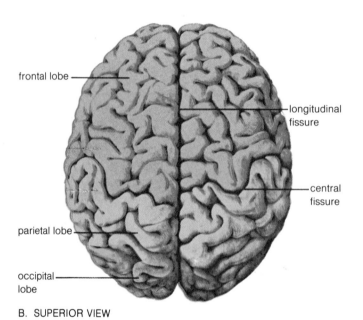

Figure 44-1 External view of the brain showing some of the fissures and convolutions (From *Human Anatomy and Physiology* by Joan G. Creager. © 1983 by Wadsworth, Inc. Reprinted by permission of Wadsworth Publishing Company, Belmont, California 94002 )

## Cerebral Functions

Each lobe of the cerebral hemispheres controls different types of functions.

1. *Frontal lobe* — The cerebral cortex of the frontal lobe controls the *motor* functions of humans. The *motor area* occupies a long band of cortex, just in front of the fissure of Rolando, in the posterior part of the frontal lobe. This motor area controls the voluntary muscles. Cells in the right hemisphere activate voluntary movements which occur in the left side of the body; the left hemisphere controls voluntary movements of the right side. The frontal lobe also includes two areas which control speech.

2. *Parietal lobe* — The parietal lobe comprises the *sensory (somesthetic)* area. It is found behind the fissure of Rolando, in front of the parietal lobe. This area receives and interprets nerve impulses from the sensory receptors for pain, touch, heat, and cold. It further helps in the determination of distances, sizes, and shapes.

3. *Occipital lobe* — The occipital lobe, located over the cerebellum, houses the *visual area,* controlling eyesight.

4. *Temporal lobe* — The upper part of the temporal lobe contains the *auditory area;* the anterior part of the lobe is occupied by the *olfactory (smell) area.*

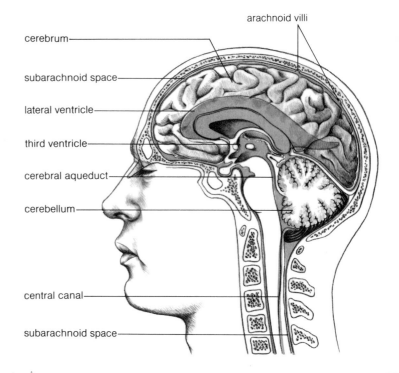

Figure 44-2 Cross section of brain, showing the arachnoid villi and subarachnoid space, as well as other structures (From *Human Anatomy and Physiology* by Joan G. Creager. © 1983 by Wadsworth, Inc. Reprinted by permission of Wadsworth Publishing Company, Belmont, California 94002.)

The cerebral cortex also controls conscious thought, judgment, memory, reasoning, and will power. This high degree of development makes the human the most intelligent of all animals.

The *diencephalon* is located between the cerebrum and the midbrain. It is composed of two major structures, the thalamus and the hypothalamus. The *thalamus* is a spherical mass of gray matter. It is found deep inside each of the cerebral hemispheres, lateral to the third ventricle. The thalamus acts as a relay station for incoming and outgoing nerve impulses. It receives direct or indirect nerve impulses from the various sense organs of the body (with the exception of olfactory sensations). These nerve impulses are then relayed to the cerebral cortex. The thalamus also receives nerve impulses from the cerebral cortex, cerebellum, and other areas of the brain. Damage to the thalamus may result in increased sensibility to pain, or total loss of consciousness.

The *hypothalamus* lies below the thalamus. It forms part of the lateral walls and floor of the third ventricle. A bundle of nerve fibers connects the hypothalamus to the posterior pituitary gland, the thalamus, and the midbrain. Eight vital functions are performed by the hypothalamus:

- **Autonomic nervous control** — Regulates the parasympathetic and sympathetic systems of the autonomic nervous system.
- **Cardiovascular control** — Controls blood pressure by regulating the constriction and dilation of blood vessels and the beating of the heart.
- **Temperature control** — Helps in the maintenance of normal body temperature (37°C or 98.6°F).
- **Appetite control** — Assists in regulating the amount of food we ingest. The "feeding center," found in the lateral hypothalamus, is stimulated by hunger "pangs," which prompt us to eat. In turn, the "satiety center" in the medial hypothalamus becomes stimulated when we have eaten enough.
- **Water balance** — Within the hypothalamus, certain cells respond to the osmotic pressure (osmolality) of the blood. When osmolality is high, due to water deficiency, the antidiuretic hormone (ADH) is secreted. ADH is produced in the hypothalamus and secreted by the posterior pituitary gland. Secreted into the bloodstream, ADH causes the kidneys to conserve water; this, in turn, keeps the blood from becoming too concentrated. A "thirst area" is found near the satiety area, becoming stimulated when the blood's osmolality is high. This causes us to consume more liquids.
- **Gastrointestinal control** — Increases intestinal peristalsis and secretion from the intestinal glands.
- **Emotional state** — Plays a role in the display of emotions such as fear and pleasure.
- **Sleep control** — Helps keep us awake when necessary.

## CEREBELLUM

The cerebellum is much smaller than the cerebrum and is also divided into two hemispheres. The function of the cerebellum is to coordinate the muscular movements of the body and thus govern the steadiness of these movements.

## BRAINSTEM

The brainstem is made up of three parts: the midbrain, pons, and the medulla. The *pons* is located in front of the cerebellum, between the midbrain and the medulla oblongata. It contains interlaced transverse and longitudinal myelinated, white nerve fibers mixed with gray matter. The pons serves as a two-way conductive pathway for nerve impulses between the cerebrum, cerebellum, and other areas of the nervous system. In this way, nerve impulses are transmitted between the two cerebellar hemispheres, from the cerebellum to the midbrain and cerebrum, and from the cerebellum to the medulla and spinal cord. The pons is also the site for the emergence of four pairs of cranial nerves, and it contains a center that controls respiration.

The *medulla oblongata* is a bulb-shaped structure found between the pons and the spinal cord. It lies inside the cranium, above the foramen magnum, in the occipital bone. The medulla is white on the outside, just like the pons, because of myelinated nerve fibers. Its functions include:

- Serving as a passageway for nerve impulses between the spinal cord and the brain
- Slowing the heart rate via the *cardiac inhibitory center*
- Controlling the rate and depth of respiration via the *respiratory center*
- Causing the dilation and constriction of blood vessels, thereby affecting blood pressure via the *vasoconstrictor center*

## SPINAL CORD

The spinal cord continues down from the medulla. It is white and soft and lies within the vertebrae of the spinal column. Like the brain, the spinal cord is submerged in cerebrospinal fluid and is surrounded by the three meninges. The gray matter in the spinal cord is located in the internal section; the white matter composes the outer part, figure 44-3. In the gray matter of the cord, connections can be made between incoming and outgoing nerve fibers which provide the basis for reflex action. The spinal cord functions as a reflex center and as a conduction pathway to and from the brain.

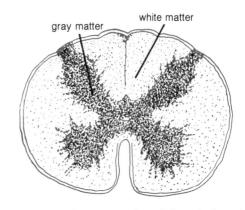

Figure 44-3 Cross section of the spinal cord

---

### Further Study and Discussion

- If laboratory facilities are available, make arrangements to observe the dissection of the brain of a sheep or calf. Locate the cerebrum, cerebellum, medulla, convolutions, hemispheres.

- From memory, draw a diagram of the brain and identify the structures. Compare it with color plate 17 upon completion.

- Make up a chart listing the parts of the brain and spinal cord. Do outside research and add functions which may not have been covered in this unit.

## Assignment

Match each term in column I with its description in column II.

| Column I | Column II |
|---|---|
| _____ 1. meninges | a. the innermost covering of brain tissue |
| _____ 2. cortex | b. separates the other two meninges |
| _____ 3. arachnoid | c. matter which contains naked nerve fibers |
| _____ 4. cerebellum | d. fluid made in the ventricles of the brain |
| _____ 5. pia mater | e. largest of 3 parts of the brain |
| _____ 6. white matter | f. nerve fibers covered with myelin sheath |
| _____ 7. cerebrum | g. part of the brain which coordinates movements |
| _____ 8. gray matter | h. the highest center of the brain |
| _____ 9. dura mater | i. the outer meninge |
| _____ 10. cerebrospinal fluid | j. regulates heartbeat, respiration, and coughing |
| | k. three membranes that cover the brain |

# Unit 45
# THE PERIPHERAL AND AUTONOMIC NERVOUS SYSTEMS

## KEY WORDS

afferent nerve
autonomic nervous system
axon
connecting neuron
dendrite
efferent nerve
effector
mixed nerve
motor neuron
peripheral nervous system
receptor
reflex act
reflex arc
sensory neuron
stimulus

## OBJECTIVES

- Relate the functions of the sympathetic and parasympathetic nervous systems
- Explain how a simple reflex act is carried out by the nervous system
- Define the terms associated with a reflex action
- Define the Key Words related to this unit of study

A nerve is composed of bundles of nerve fibers and small vessels all enclosed by connective tissue. If the nerve is composed of fibers that carry impulses from the sense organs to the brain or spinal cord, it is called a *sensory,* or *afferent,* nerve; if it is composed of fibers carrying impulses from the brain or spinal cord to muscles or glands, it is known as a *motor,* or *efferent,* nerve; and if it contains both sensory and motor fibers, it is referred to as a *mixed nerve.*

Certain of the twelve pairs of cranial nerves (such as the facial nerve) are mixed nerves; some (as the optic nerve) contain only sensory fibers; and others are entirely motor nerves. All thirty-one pairs of spinal nerves are mixed nerves. Each of them divides into branches. These go either directly to a particular body segment, or they form networks with adjacent spinal nerves known as *plexuses.*

## PERIPHERAL NERVOUS SYSTEM

The peripheral nervous system includes all the nerves of the body, see figure 45-1. It connects the central nervous system to the various body structures. The autonomic nervous system is a specialized part of the peripheral system; it controls the involuntary, or automatic, activities of the vital internal organs.

## AUTONOMIC NERVOUS SYSTEM

The autonomic nervous system includes nerves, ganglia, and plexuses which carry

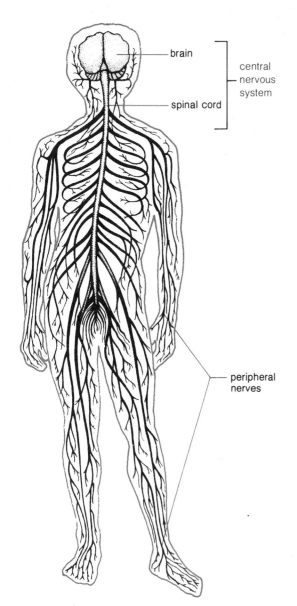

Figure 45-1 The peripheral nervous system connects the central nervous system to the various structures of the body. Messages are relayed from these structures back to the brain via the spinal cord. Spinal nerves leave the spinal cord through openings in the vertebrae, extending, dividing, and branching out into an intricate and complicated network reaching every structure of the body. Twenty-five pairs of spinal nerves pass between the vertebrae and branch out to all parts of the body. *(Courtesy of CPR, Teaneck, N.J.)*

impulses to all smooth muscle, secretory glands, and heart muscle, figure 45-2. It thus regulates the activities of the *visceral organs* (heart and blood vessels, respiratory organs, alimentary canal, kidneys and urinary bladder, and reproductive organs). The activities of these organs are usually automatic and not subject to conscious control.

The autonomic system is comprised of two divisions: the *sympathetic* and the *parasympathetic*. These two divisions are antagonistic in their action. For example, the sympathetic system may accelerate the heartbeat in response to fear, whereas the parasympathetic slows it down. Normally the two divisions are in balance; the activity of one or the other becomes dominant as dictated by the needs of the organism.

The sympathetic nervous system consists primarily of two cords, beginning at the base of the brain and proceeding down both sides of the spinal column. These are made up of nerve fibers and ganglia of nerve cell bodies. The cord between the ganglia is a cable of nerve fibers, closely associated with the spinal cord. Sympathetic nerves extend to all the vital internal organs, including the liver and pancreas, as well as innervating the heart, stomach, intestines, blood vessels, the iris of the eye, sweat glands, and the bladder.

The parasympathetic system is composed of two important active nerves: the vagus and the pelvic nerves. The vagus nerve, which extends from the medulla and proceeds down the neck, sends branches to the chest and neck. The pelvic nerve, emerging from the spinal cord around the hip region, sends branches to the organs in the lower part of the body.

Both the sympathetic and parasympathetic are strongly influenced by emotion. During periods of fear, anger, or stress, the

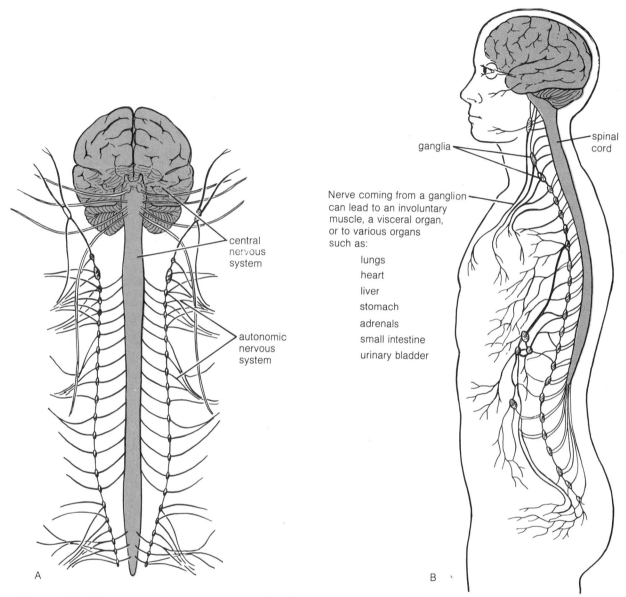

Figure 45-2 The autonomic nervous system (involuntary nervous system) governs those functions which are carried out automatically, without conscious thought, such as regulation of heart beat, digestion of food, etc.

sympathetic division acts to prepare the body for action. The effects of the parasympathetic are generally to counteract the effects of the sympathetic. For example, the sympathetic system constricts the blood vessels, thus raising blood pressure, while the parasympathetic system dilates the blood vessels, lowering blood pressure. The two systems operate as a pair, striking a nearly perfect balance when the body is functioning properly.

## The Reflex Act

The simplest type of nervous response is the *reflex* act, which is unconscious and involuntary. The blinking of the eye when a particle of dust touches it, the removing of the finger from a hot object, the secretion of saliva at the sight or smell of food, the movements of the heart, stomach, and intestines, are all examples of reflex actions.

Every reflex act is preceded by a stimulus. Anything in the environment which causes activity is called a *stimulus*. Examples of stimuli are sound waves, light waves, heat energy, and odors. Special structures called *receptors* pick up these stimuli. For example, the retina of the eye is the receptor for light; special cells in the inner ear are receptors for sound waves; and special structures in the skin are the receptors for heat and cold.

Reaction to a stimulus is called the *response*. The response may be in the form of movement; in which case, the muscles are the *effectors*, or responding organs. If the response is in the form of a secretion, the glands are the effectors. Reflex actions involving the skeletal muscles are controlled by the spinal cord. They involve only sensory, connecting, and motor neurons. This pathway is known as the *reflex arc*, figure 45-3.

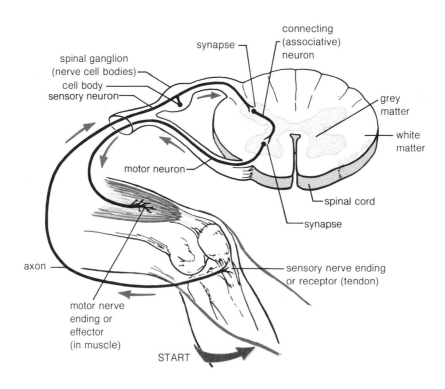

Figure 45-3 The reflex arc. In this example, tapping the knee (patellar tendon) results in extension of the leg, producing the knee jerk or reflex.

## Further Study and Discussion

- Discuss the action of the autonomic nervous system when a person is served an appetizing lunch.
- List ten reflex acts.

## Assignment

A. Complete the following statements.

1. A nerve is composed of small blood vessels and of bundles of fibers called _____ .

2. A nerve composed of fibers carrying impulses from sense organs to the brain or spinal cord is called a _____ or _____ nerve.

3. A nerve composed of fibers which carry impulses from the brain or spinal cord to muscles or glands is called a _____ or _____ nerve.

4. A mixed nerve contains both _____ and _____ fibers.

5. The autonomic nervous system is a specialized part of the peripheral system and controls _____ .

6. The autonomic nervous system has two parts which counterbalance each other; these are the _____ and _____ systems.

B. Identify the structures on the following diagram. Enter your answers below it.

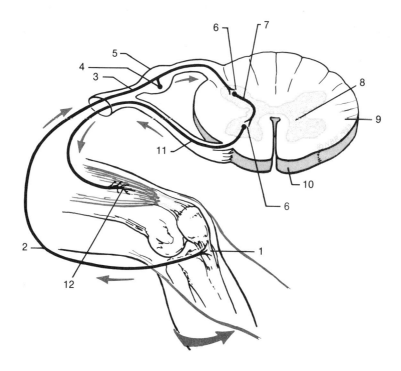

① _____   ⑦ _____

② _____   ⑧ _____

③ _____   ⑨ _____

④ _____   ⑩ _____

⑤ _____   ⑪ _____

⑥ _____   ⑫ _____

# Unit 46
# SPECIAL SENSE ORGANS: THE EYE AND EAR

## KEY WORDS

anterior chamber
anvil
choroid coat
cochlea
cones
cornea
dilator pupillae
eustachian tube
extrinsic muscles
hammer
intrinsic muscles
iris
lens
orbital socket
organ of Corti
posterior chamber
pupil
retina
rods
sclera
semicircular canals
sphincter
stirrup
suspensory ligament
vitreous humor

## OBJECTIVES

- Explain how stimulation of a sense organ results in sensation
- Identify the parts of the eye, and relate them to their function
- List the parts of the ear and relate them to their function
- Define the Key Words that relate to this unit of study

Sensory receptors are special structures which are stimulated by changes in the environment. Sensory receptors (touch, pain, temperature, and pressure) are found all over the body, located either in the skin or connective tissues. Special sensory receptors include the taste buds of the tongue, special cells in the nose, the retina of the eye, and the special cells in the inner ear which make up the organ of Corti. When a sense organ is stimulated, the impulse travels along nerve pathways to the brain, where it is registered in a certain area. Sensation actually takes place in the brain, but it is mentally referred back to the sense organ. This is called *projection* of the sensation.

## THE EYE

The human eye is a tender sphere about 1″ in diameter (about 2.5 cm). It is protected by the *orbital socket* of the skull. The eyeball is moved by muscles. The eye is protected by the bone surrounding it and by the eyebrows, eyelids, and eyelashes, figure 46-1.

The location of the eyes in front of the head allows for superimposition of images from each eye. This enables us to see stereoscopically, in three dimensions (length, width, and depth). The eye's optical system for detecting light is similar to that of a camera.

The wall of the eye is made up of three concentric layers, or coats, each with its specific function. These three layers are the sclera, the choroid, and the retina, see color plate 18.

### Sclera

The outer layer is called the *sclera*, or white of the eye. It is a tough, unyielding, fibrous capsule which maintains the shape of

**282** SECTION 10 COORDINATION OF BODY FUNCTIONS

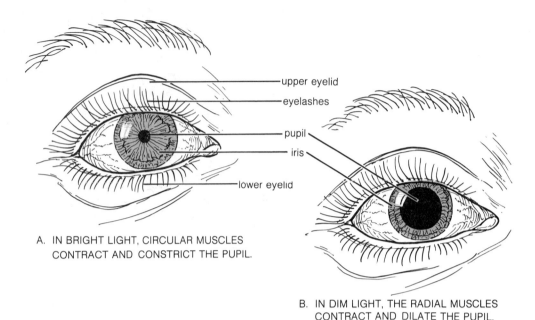

A. IN BRIGHT LIGHT, CIRCULAR MUSCLES CONTRACT AND CONSTRICT THE PUPIL.

B. IN DIM LIGHT, THE RADIAL MUSCLES CONTRACT AND DILATE THE PUPIL.

Figure 46-1 External view of the eye

the eye and protects the delicate structures within. Muscles responsible for moving the eye within the orbital socket are attached to the outside of the sclera. These muscles are referred to as the *extrinsic muscles*. They include the superior, inferior, lateral, medial rectus, and the superior and inferior oblique. See table 46-1 for a listing of the extrinsic eye muscles and their functions.

## Cornea

In the very front center of the sclerotic coat lies a circular, clear area called the *cornea*. The cornea is sometimes referred to as the "window" of the eye. It is transparent to permit light rays to pass through it. This transparency is due to the lack of blood vessels. Thus corneal cells are fed by the movement of lymph through interstitial, or lymph spaces. The cornea is composed of five layers of flat cells arranged much like sheets of plate glass. Possessing pain and touch receptors, it is sensitive to any foreign particles that come in contact with its surface. An injury to the cornea causes scarring and impaired vision.

Table 46-1 Extrinsic and Intrinsic Eye Muscles

| EYE MUSCLE | FUNCTION |
| --- | --- |
| A. Extrinsic | |
| 1. Superior rectus | Rolls eyeball upward |
| 2. Inferior rectus | Rolls eyeball downward |
| 3. Lateral rectus | Rolls eyeball laterally |
| 4. Medial rectus | Rolls eyeball medially |
| 5. Superior oblique | Rolls eyeball on its axis, moves cornea downward and laterally |
| 6. Inferior oblique | Rolls eyeball on its axis, moves cornea upward and laterally |
| B. Intrinsic | |
| 1. Sphincter pupillae | Constricts pupil |
| 2. Dilator pupillae | Dilates pupil |

## Choroid Coat and the Iris

The middle layer of the eye is the *choroid coat*. It contains blood vessels to nourish the eye, and a non-reflective pigment rendering it dark and opaque. The pigment provides the choroid coat with a deep, red-purple color; this darkens the eye chamber, preventing light reflection within the eye. In front, the choroid coat has a circular opening called the *pupil*. A colored, muscular layer surrounds the pupil; this is the *iris,* or colored part of the eye. The iris may be blue, green, gray, brown, or black. Eye color is related to the number and size of melanin pigment cells in the iris. If there is little melanin present, the eye is blue, because light is scattered to a greater extent. With increasing quantities of melanin, eye color ranges from green to black. The total absence of melanin results in a pink eye color, characteristic of albinism. Such irises are pink because the blood inside the choroid blood vessels shows through the iris.

Within the iris are two sets of antagonistic, smooth muscles, the *sphincter* and the *dilator pupillae*. These muscles help the iris to control amounts of light entering the pupil. When the eye is focused on a close object or stimulated by bright light, the sphincter pupillae muscle contracts, rendering the pupil smaller. Conversely, when the eye is focused on a distant object or stimulated by dim light, the dilator pupillae muscle contracts. This causes the pupil to grow larger, permitting as much light as possible to enter the eye. In this way the eye may be compared to a camera; the iris corresponds to the shutter or diaphragm.

## Lens and Related Structures

The *lens* is a crystalline structure located behind the iris and pupil. It is composed of concentric layers of fibers and crystal-clear proteins in solution. It is an elastic, disc-shaped structure with anterior and posterior convex surfaces, thus forming a biconvex lens. However, the posterior surface is more curved than that of the anterior. The curvature of each surface alters with age. During infancy, the lens is spherical; in adulthood, medium convexed; and almost flattened in old age. The capsule surrounding the lens also loses its elasticity over a period of time. The lens is held in place behind the pupil by *suspensory ligaments* from the *ciliary body* of the choroid coat.

The lens is situated between the *anterior* and *posterior chambers*. The anterior chamber is filled with a watery fluid referred to as the *aqueous humor,* and it is constantly replenished by blood vessels behind the iris. *Vitreous humor,* a transparent jellylike substance, fills the posterior chamber. Both of these substances help to maintain the eyeball's spherical shape, refracting (bending) light rays as they pass through the eye.

## Retina

The *retina* of the eye is the innermost, or third coat of the eye. It is located between the posterior chamber and the choroid coat. The retina does not extend around the front portion of the eye. It is upon this light-sensitive layer that light rays from an image are formed. After the image is focused on the retina, it travels via the optic nerve to the visual part of the cerebral cortex. This is found towards the back of the head, just above the neck. If light rays do not focus correctly on the retina, the condition may be corrected with properly fitted contact lenses, or eyeglasses, which bend the light rays as required.

1. Close your left eye and focus your right eye on the cross.
2. Move the page slowly away from your eye and then slowly toward your eye.
3. At a distance of about 6-8 inches the black circle "disappears."

**Figure 46-2 Testing for the blind spot**

The retina contains pigment and specialized cells known as *rods* and *cones* which are sensitive to light. The part of the retina where the nerve fibers enter the optic nerve to go to the brain does not have these specialized cells; therefore, it is most sensitive to light. For this reason, it is often called the "blind spot."

**The Optic Disc and the Fovea.** Viewing the retina through an ophthalmoscope, one can observe a yellow disc known as the *macula lutea*. Within this disc is the fovea centralis, which contains the cones for color vision, see color plate 18. The area around the fovea centralis is the *extrafoveal* or *peripheral region*. This is where the rods for dim and peripheral vision can be found.

Slightly to the side of the fovea lies a pale disc called the *optic disc* or *blind spot*. Nerve fibers from the retina gather here to form the nerve. The optic disc contains no rods or cones; therefore, it is devoid of visual reception.

See figure 46-2 to help you locate your blind spot.

# THE EAR

The ear is a special sense organ which is especially adapted to pick up sound waves and send these impulses to the auditory center of the brain, located in the temporal area just above the ears. The receptor for hearing is the delicate *organ of Corti*, which is located within the cochlea of the inner ear.

The ear has three parts: the outer or external ear, the middle ear, and the inner ear, see figure 46-3. The outer ear consists of the visible portion and a canal which leads to the ear drum (*tympanic membrane*).

The middle ear is really the cavity in the temporal bone. It connects with the pharynx by means of a tube called the *eustachian tube*. This tube serves to equalize the air pressure in the middle ear with that of the outside atmosphere. A chain of three tiny bones is found in the middle ear: the *hammer*, the *anvil*, and the *stirrup*; they transmit sound waves from the ear drum to the inner ear.

The inner ear consists of several membrane-lined channels which lie deep within the temporal bone. The special organ of hearing is a spiral-shaped passage known as the *cochlea*, which contains a membranous tube called the *cochlear duct*. The duct is filled with fluid that vibrates when the sound waves from the stirrup bone strike against it. Located in the cochlear duct are delicate cells which make up the organ of Corti. These hairlike cells pick up the vibrations caused by sound waves against the fluid, then they transmit them through the auditory nerve to the hearing center of the brain.

Three *semicircular canals* also lie within the inner ear. They contain a liquid, and delicate hairlike cells which bend when the liquid is set in motion by head and body movements. These impulses are sent to the brain, helping to maintain body balance, or equilibrium. They have nothing to do with the sense of hearing.

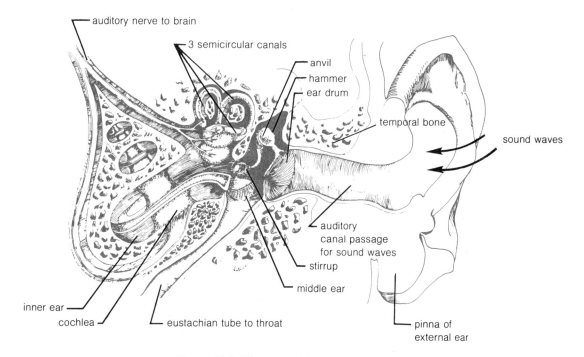

Figure 46-3 The ear and its structures

## Further Study and Discussion

- Using an eye model, identify the chief structures in the eye. Explain their functions.

- Using an ear model, identify the chief structures in the ear. Explain their functions.

- Using a diagram or a chart of the skin, identify the four receptors found there.

- Locate the taste buds on the tongue, using a chart.

- Using color plate 18, trace the path of light rays from the time they strike the cornea until the sensation of sight is registered in the brain.

- In figure 46-3, trace the path of sound waves from the time they strike the ear drum until the sensation of hearing is registered in the brain.

**286** SECTION 10 COORDINATION OF BODY FUNCTIONS

## Assignment

1. Name the three structural coats found in the eye. Give the function of each one.

2. Explain how refraction of light is important to vision.

3. Describe how the organ of Corti functions.

4. Identify the structures in the following sketch and write your answers in the spaces provided.

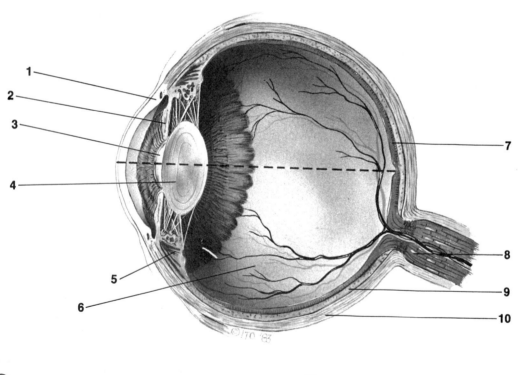

① _____
② _____
③ _____
④ _____
⑤ _____

⑥ _____
⑦ _____
⑧ _____
⑨ _____
⑩ _____

## KEY WORDS

anticonvulsant
arbovirus
herpetic neuralgia
miotic
paresthesia
spastic quadriplegia
topically
unilateral

# Unit 47
# REPRESENTATIVE DISORDERS OF THE NERVOUS SYSTEM

## OBJECTIVES

- Recognize symptoms of some common ailments of the nervous system
- Relate treatment to some common nervous system diseases
- List common ear and eye disorders and related treatment
- Identify the Key Words in this unit of study

The health care provider must become familiar with symptoms and treatments for disorders of the nervous system. Some of the more common ones are mentioned here.

*Chorea* is a nervous system disorder characterized by involuntary muscular twisting and writhing movements of the arms, face, and legs. The symptoms frequently appear following streptococcal infection. The disease may last from three to six months. Treatment consists of rest, nourishing food, and avoidance of excitement of any kind. The disorder is also known as *chorea minor, dancing chorea, St. Vitus' dance,* or *Sydenham's chorea.*

*Shingles* or *herpes zoster* is an acute viral nerve infection. It is characterized by a unilateral (one-sided) inflammation of cutaneous nerves. Although intercostal nerves are commonly affected, the course of nerve inflammation can spread virtually to any nerve. Symptoms include extremely painful vesicular eruptions of the skin and mucous membranes along the route of the inflamed nerve. The extreme pain resulting from shingles is known as *herpetic neuralgia.* Shingles is often found in elderly or debilitated persons. The afflicted area, typically the chest or abdomen, must be treated by protecting it from air and from clothing irritation.

*Neuralgia* is a sudden severe, sharp, stabbing pain along the pathway of a nerve. The pain is often brief, usually a symptom of a disease. The various forms of neuralgia are named according to the nerve that they affect. For instance, neuralgia of the trigeminal nerve (sensory nerve of the face and head) is called *trigeminal neuralgia.* The pain from trigeminal neuralgia and other forms of neuralgia can be relieved with analgesics and/or narcotics.

*Neuritis* is the inflammation of a nerve trunk. This condition can cause extreme pain, hypersensitivity, loss of sensation, muscular atrophy, paralysis, weakness and *paresthesia* (tingling, burning and crawling of the skin). The different types of neuritis are named according to their cause, location and pathology. They include four major types:

1. *alcoholic neuritis* — neuritis deriving from chronic alcoholism, presumably a result of thiamine deficiency because of insufficient and improper diet.

2. *ascending neuritis* — neuritis that extends from the periphery of a nerve to the spinal cord or brain.

3. *infectious neuritis* — an acute multiple neuritis caused by a viral disease.

4. *sciatic neuritis or sciatica* — this is a form of neuralgia or neuritis of the sciatic nerve of the body.

The pain from these various forms of neuritis can be relieved by administering pain killers or analgesics.

*Poliomyelitis,* or *infantile paralysis,* is a viral infectious disease of the nerve pathways in the spinal cord. The muscles which are controlled by these diseased nerve paths become paralyzed. Death may result. Treatment consists of hot packs, exercises given by a trained professional, and special exercises given under water. The patient may have to be placed in an "iron lung" if the muscles of respiration are involved. The Sister Kenny method of treatment is administered by trained specialists. Vaccines (Salk and Sabin) are now available to protect against polio, and all children should be immunized against this crippling disease.

*Encephalitis* is a viral disease characterized by inflammation of the brain. There are several causes of encephalitis in addition to viruses, including various other organisms and certain chemical substances. A number of viruses have been isolated and identified as the major causative factors. Encephalitis results primarily from the bite of a mosquito carrying the encephalitis virus; hence the term *arbovirus* (arthropod-borne virus). The different forms of encephalitis are all characterized by fever, lethargy, extreme weakness, and visual disturbances. No adequate preventive immunization methods have yet been developed, nor is there a specific treatment for the infections. A strain known as **California encephalitis** has been reported throughout the United States. This severe disease, which primarily affects children, is accompanied by coma and convulsions; fortunately the prognosis is excellent.

*Cerebral palsy* is a disturbance in voluntary muscular action due to brain damage. Definite causes are unknown. It may result from birth injuries, intracranial hemorrhage, or infections such as encephalitis. The most pronounced characteristic of cerebral palsy is *spastic quadriplegia.* Spastic quadriplegia involves spastic paralysis of all four limbs, though it is most pronounced in the legs.

*Hydrocephalus* is an increased volume of cerebrospinal fluid within the cavity of the brain. This causes enlargement of the skull and prominence of the forehead.

*Convulsions,* characterized by violent muscle contractions, may occur because of high fever, lack of vitamin D, or brain tumors; they cause brain tissue to discharge abnormal nerve signals.

*Acute bacterial meningitis* is the inflammation of the membranes, or meninges, of the brain or spinal cord. The different types of meningitis are classified according to their causative organisms: tuberculosis meningitis, pneumococcal meningitis, or meningococcus meningitis. The latter, meningococcus meningitis, is caused by **Neisseria meningitidis**.

Epidemics of this disease occur infrequently, although isolated cases do come up. It is endemic to such close-living groups as residents of college dormitories and members of the armed services. These groups may suffer frequent outbreaks due to the introduction of new carriers, or susceptibles. In the military, these outbreaks range from 200 to 500 cases a year.

Symptoms of meningococcus meningitis include headache, fever, nasal secretions, sore throat, back and neck pain, loss of mental alertness, and rashes. Treatment by intravenous injection of immune serum (or injection directly into the spinal canal) has met with limited success. Immunity of unpredictable duration follows recovery from the inflammation. Chemotherapy with penicillin, rifampin, sulfanomides, sulfadiazine, and other antibiotics is fairly successful. However, even this method of treatment has its limitations due to the possible development of drug-resistant strains. Another problem is that drugs do not readily pass through the meninges to reach the invading bacteria.

*Epilepsy* is a disorder of the brain, characterized by a recurring and excessive electrical discharge from neurons. Approximately one person out of 200 in the United States suffers from some form of epilepsy. Epileptic seizures are believed to be a result of spontaneous, uncontrolled, reverberating cycles of electrical activity in the neurons of the brain. One portion of the brain stimulates another, setting off a cycle of activity that accelerates and runs its course until the neurons become fatigued. When such electrical reverberations occur, the subject may suffer hallucinations and a seizure. The confused neuronal electrical circuitry leads to a loss of consciousness; the neuronal fatigue induces sleep. *Grand mal,* or severe seizure, is less frequent than the *petit mal* (milder seizure). In petit mal, some of the victims see flashes of light. There may be odor and sound sensations, even a blackout of consciousness. Most epileptic persons can lead normal lives with regular medication.

Medications used to control seizures are referred to as *anticonvulsants.* Diphenylhydantoin (Dilantin) helps to control grand mal seizures. Trimethadione (Tridione) and paramethadione (paradione) control petit mal.

## EAR DISORDERS

*Otitis media* is an infection of the middle ear. It usually causes earache. This disorder is often a complication of the common cold in children. Treatment with antibiotics will cure the infection.

## EYE DISORDERS

*Conjunctivitis* is an inflammation of the conjunctival membranes in front of the eye. Redness and discharge of mucus occur. Since it may be contagious, conjunctivitis should be promptly treated by a physician. Treatment includes eye washes, or irrigations, to flush the conjunctiva. Eye irrigations cleanse and relieve the inflammation and the pain. These solutions usually consist of weak solutions of normal isotonic saline and/or boric acid.

*Glaucoma* is an eye condition in which the aqueous humor does not circulate properly within the eye and pressure increases within the eyeball. If untreated, glaucoma leads to blindness because it damages the retina and optic nerve. With prompt treatment, total blindness may be avoided. Early detection is most important, and treatment usually prevents progress of the disease.

Treatment involves the use of *miotics* (drugs causing constriction of the pupil). Miotic drugs lower eyeball pressure. These include pilocarpine nitrate (Pilocarpine Hydrochloride), the effects of which last only several hours, and demercarium bromide (Humorsol), with effects that last from 5 to 10 days.

A *cataract* is a condition characterized by a lack of transparency of the lens. Light cannot pass through the clouded lens; therefore the person cannot see. The vision is corrected with eyeglasses or surgery; the opaque lens is removed and replaced by a synthetic lens.

A *sty* is a tiny abscess at the base of an eyelash. It is due to inflammation of one of the sebaceous glands of the eyelid. Various antibiotics can be used as a curative. Bacitracin is especially useful as an ophthalmic anti-infective, because it is non-irritating to the eye and produces no known side effects. Few microorganisms are able to develop a resistance to bacitracin. It is administered topically in an ophthalmic ointment.

## Further Study and Discussion

- Discuss the procedure that should be followed in caring for a patient seized by an epileptic attack. What observations should be reported?
- Prepare for a panel discussion on rehabilitation of cerebral palsy patients. Include family and social problems.
- Explain how deafness can be helped or treated.
- Discuss the reason for the infrequency of poliomyelitis today.

## Assignment

A. Match each of the terms in column I with its description in column II.

| Column I | Column II |
| --- | --- |
| _____ 1. acute bacterial meningitis | a. disease which may be characterized by convulsions |
| _____ 2. poliomyelitis | b. infection of the brain membranes |
| _____ 3. epilepsy | c. middle ear infection |
| _____ 4. glaucoma | d. viral disease for which there is now a vaccine |
| _____ 5. cataract | e. a clouded or opaque lens |
| | f. condition caused by increased pressure within the eyeball |

B. Answer the following questions.

1. What is the difference between neuralgia and neuritis?

2. What is sciatica?

3. What disease results when the membranes of the brain become inflamed?

4. What are the hazards of untreated glaucoma?

# SELF-EVALUATION

## Section 10
## COORDINATION OF BODY FUNCTIONS

A. Identify each numbered part and give its function.

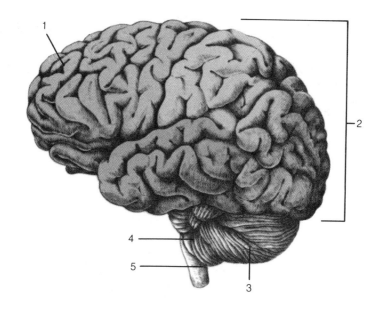

1 _____

2 _____

3 _____

4 _____

5 _____

B. Using the diagram of the reflex arc, trace the path of the impulse. Identify the numbered parts and describe the action.

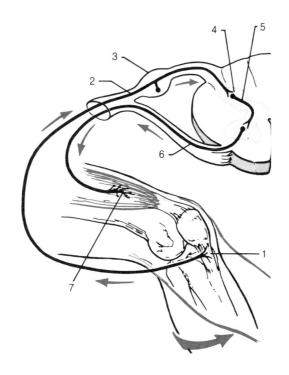

① _____
② _____
③ _____
④ _____
⑤ _____
⑥ _____
⑦ _____

C. Place the correct answers in the blank spaces.
1. The unit of structure in nervous tissue is the _____.
2. There are two chief types of neurons, the _____ or _____ which carries messages to the brain and the _____ or _____ which convey messages away from the central nervous system to peripheral areas.
3. Axons of nerve cells have a protective covering around them called the _____.
4. The brain and spinal cord are covered by three _____. They are, from the center out, _____, _____, and _____.
5. The three main divisions of the brain include the _____, _____, and _____.
6. The spinal cord lies within the _____.
7. Connections between the brain and surrounding structures are established by _____ nerves.
8. Connections between the spinal cord and other structures are established by _____ nerves.
9. The _____ is the largest part of the human brain.
10. The four receptors for sensation which are found in the skin are those for touch, cold, heat, and _____.

# Glossary

## A

**abdomen** (ab′-do-mun): portion of body lying between thorax and pelvis

**abdominal hernia** (ab-dom′-i-nul hur′-nee-uh): abnormal protrusion of an organ, or part of an organ, through abdominal wall

**abduction** (ab-duck′-shun): movement away from midline or axis of body

**absolute zero** (ab′-suh-lewt zee′-row): complete absence of heat or about −273.2° C (−459.8°F)

**absorption** (ub-sorp′-shun): passing of a substance into body fluids and tissues

**accessory digestive organs** (ack-sez′-uh-ree di-jes′-tiv or′-gunz): structures helping in mechanical digestion of food; glands producing secretion to assist chemical digestion in mouth, stomach, small intestine. Accessory organs include teeth, tongue, salivary glands, pancreas, liver, and gallbladder

**acetabulum** (as″-e-tab′-yoo-lum): a cup-shaped cavity in the innominate bone receiving the head of the femur

**Achilles tendon** (ack-i-′-leez ten′-dun): cord at rear of heel

**acid** (as′-id): chemical compound that ionizes to form hydrogen ions (H⁺) in aqueous solution

**acidosis** (as″-i-do′-sis): disturbance in the acid-base balance from excess acid, or excessive loss of bicarbonate; depletion of alkaline reserve

**active transport** (ack′-tiv tranz′-port): process by which solute molecules are transported across a membrane against a concentration gradient, from an area of low concentration to one of high concentration

**adduction** (a-duck′-shun): movement of part of body or limb toward the midline of body; opposite of abduction

**adenoids** (ad′-e-noydz): pair of glands composed of lymphoid tissue, found in nasopharynx; also called *pharyngeal tonsils*

**adenosine diphosphate (ADP)** (a-den′-o-seen dye-fos′-fate): chemical compound consisting of one molecule of adenine, one of ribose, two of phosphoric acid. Intermediary compound in the production of energy for cellular and muscular activity

**adenosine triphosphate (ATP)** (a-den′-o-seen try-fos′-fate) chemical compound consisting of one molecule of adenine, one of ribose, three of phosphoric acid

**adipose** (ad′-i-pose): fatty or fat-like

**afferent nerve** (af′-ur-unt nurv): a nerve that carries nerve impulses from the periphery to the central nervous system; also known as a *sensory nerve*

**agglutinin** (a-gloo′-ti-nin): antibody found in normal or immune serum, causing antigen and cellular clumping

**agglutinogen** (a-gloo′-tin-o-jen): chemical substance (antigen) which stimulates the formation of a specific agglutinin

**agranulocyte** (ay-gran′-yoo-lo-site): nongranular, white blood cell; known as *agranular leukocyte*

**albinism** (al'-bin-izm): partial or total absence of melanin pigment from eyes, hair, and skin

**albuminuria** (al-bew"-mi-new'-ree-uh): excess of albumin protein in urine

**alimentary canal** (al"'-i-men'-tuh-ree kuh-nal'): entire digestive tube from mouth (ingestion) to anus (excretion)

**alkalosis** (al"-kuh-lo'-sis): excessive alkali; disturbance in acid-base balance from excess loss of acid

**alveoli** (al-vee'-o-li): air cells found in lung

**ameboid** (a-mee'-boyd): resembling amoeba in form or movement

**amenorrhea** (a-men"-o-ree'-uh): absence of menstruation

**anal sphincter** (ay'-nul sfink'-tur): muscles surrounding anal opening

**anemia** (uh-nee'-mee-uh): blood disorder characterized by reduction in number of red blood cells or hemoglobin

**aneurysm** (an'-yoo-rizm): a widening, or sac, formed by dilation of a blood vessel

**angina pectoris** (an-ji'-nuh peck'-to-ris): attacks of tightness, choking or suffocation, often accompanied by severe chest pain radiating to left shoulder and arm; caused by inadequate blood and oxygen supply to heart

**ankylosis** (ank"-i-lo'-sis): abnormal immobility and consolidation of a joint

**anorexia** (an"-o-rek'-see-uh): loss of appetite

**anterior chamber** (an-teer'-ee-ur chame'-bur): space between cornea and iris

**antibody** (an'-ti-bod"-ee): substance produced by the body, that inactivates a specific foreign substance which has entered the body

**anticoagulant** (an"-tee-ko-ag'-yoo-lunt): chemical substance that prevents or slows down blood clotting (like heparin)

**anticonvulsant** (an"-tee-kun-vul'-sunt): therapeutic agent that stops or prevents convulsions

**antidiuretic hormone** (an"-tee-dye-yoo-ret'-ik hor'-mone): hormone secreted by the posterior pituitary gland, which prevents or suppresses urine excretion

**antigen** (an'-ti-jin): substance stimulating formation of antibodies against itself

**antiprothrombin** (an"-tee-pro-throm'-bin): chemical substance that directly or indirectly reduces or retards action of prothrombin (such as heparin)

**antithromboplastin** (an"-tee-throm'-bo-plas'-tin): chemical substance inhibiting clot-accelerating effect of thromboplastins

**anus** (ay'-nus): outlet from rectum

**anvil** (an'-vil): middle ear bone, or ossicle, in a chain of three ossicles of the middle ear

**aorta** (ay-or'-tuh): largest artery in body, rising from left ventricle of the heart; its branches distribute oxygenated blood to all parts of the body

**aortic semilunar valve** (ay-or'-tik sem"-ee-loo'-nur valv): made up of three half-moon-shaped cups, located between junction of aorta and left ventricle of heart

**apex** (ay'-peks): top of object; point or extremity of a cone

**apex of lung** (ay'-peks of the lung): upper extremity of lung, behind border of the first rib

**aphasia** (a-fay'-zhuh): loss of ability to speak, may be accompanied by loss of verbal comprehension

**aponeurosis** (ap"-o-new-ro'-sis): flattened sheet of white, fibrous connective tissue; serves as attachment for flat muscles, or as sheet enclosing/binding muscle groups

**appendicitis** (a-pen"-di-si-tis): inflammation of appendix

**appendicular skeleton** (ap"-en-dik'-yoo-lur skel'-uh-tun): part of skeleton consisting of pectoral and pelvic girdles, and limbs

**arachnoid** (uh-rak'-noyd): weblike middle membrane of meninges

**arbovirus** (ahr"-bo-vye'-rus): any of over 200 arthropod-borne viruses transmitted to susceptible vertebrate hosts by blood-sucking arthropods

**areola** (a-ree-o'-luh): pigmented ring around nipple or any small space in tissue

**arrhythmia** (a-rith'-mee-uh): absence of a normal rhythm in heartbeat

**arteriole** (ahr-teer'-ee-ole): small branch of artery

**arteriosclerosis** (ahr-teer"-ee-o-skleh-ro'-sis): hardening of arteries, resulting in thickening of walls and loss of elasticity

**artery** (ahr'-tur-ee): blood vessel which carries blood away from heart

**arthritis** (ahr-thry'-tis): inflammation of a joint

**ascites** (a-si'-teez): accumulation of fluid in the peritoneal cavity

**assimilation** (a-sim″-i-lay′-shun): process of changing food into form suitable for absorption by the circulatory system, and subsequent transformation into body tissue

**ataxia** (ay-tak′-see-uh): muscle incoordination, particularly of muscle groups involved in walking or reaching for objects

**atlas** (at′-lus): first cervical vertebra; articulates with axis and occipital skull bone

**atrial fibrillation** (ay′-tree-ul figh″-bri-lay′-shun): cardiac arrhythmia, characterized by rapid, irregular atrial impulses and ineffective atrial contractions; heartbeat varies from 60 to 180 per minute and is highly irregular in rhythm and intensity

**atrioventricular or A-V node** (ay″-tree-o-ven-trik′-yoo-lur node): small mass of interwoven conducting tissue underneath right atrial endocardium

**atrioventricular valves** (ay″-tree-o-ven-trik′-yoo-lur valvz): mitral (bicuspid) and tricuspid valves of heart

**atrium** (ay′-tree-um): upper chamber of heart

**atrophy** (a′-truh-fee): wasting away of tissue

**auricle** (aw′-ri-kul): (1) pinna, or ear flap of external ear; (2) atrium of the heart

**autonomic** (aw″-tuh-nom′-ik): independent or self-regulating

**autonomic nervous system** (aw″-tuh-nom′-ik nur′-vus sis′-tum): collection of nerves, ganglia, and plexuses through which visceral organs, heart, blood vessels, glands, and smooth (involuntary) muscles receive their innervation

**autosome** (aw′-to-sohm): non-sex determining chromosome

**axial skeleton** (ak′-see-ul skel′-e-tun): skeleton of head and trunk

**axilla** (ak-sil′-uh): armpit

**axis** (ack′-sis): (1) imaginary line passing through center of the body; (2) second cervical vertebra

**axon** (ack′s-on): nerve-cell structure which carries impulses away from cell body to dendrites

## B

**base** (bays): (1) lowest part of a body; (2) main ingredient of a substance; (3) chemical compound yielding hydroxyl ions (OH$^-$) in an aqueous solution which will react with acid to form a salt and water

**basophil(e)** (bay′-suh-fil): substance, cell, or tissue element showing affinity for basic dyes

**benign** (be-nine′): non-malignant

**bicarbonate ion** (by-kahr′-buh-nate eye′-on): salt of carbonic acid characterized by ion HCO$_3-$

**biceps** (bye′-seps): large flexor muscle of upper arm or leg

**bicuspid** (bye-kus′-pid). having two cusps

**bicuspid or mitral valve** (bye-kus′-pid [my′-trul] valv): atrioventricular valve of left side of heart

**bifurcation** (bye″-fur-kay′-shun): division into two branches

**bilateral symmetry** (bye-lat′-ur-ul sim′-e-tree): relating to both sides of body

**biopsy** (bye′-op-see): excision of a piece of tissue from a living body for diagnostic study

**bipedal** (bye-ped′-ul): having two feet

**bolus** (bo′-lus): rounded mass of food prepared by mouth for swallowing

**bowel** (bow′-ul): intestine

**brachial** (bray′-kee-ul): pertaining to the upper arm

**brachiocephalic artery** (bray′-kee-o-se-fal′-ik ahr′-tur-ee): artery rising from right side of aortic arch and dividing into right subclavian and right common carotid arteries

**brainstem** (brayn′-stem): portion of brain other than cerebral hemispheres and cerebellum

**bronchiole** (bronk′-ee-ole): one of small subdivisions of a bronchus (1 mm or less)

**bronchiolectasis** (bronk″-ee-o-lek′-tuh-sis): condition in which the bronchi are dilated

**bronchus** (bronk′-us): one of two primary branches of trachea

**buccal** (buck′-ul): pertaining to the cheek or mouth

**buccal (oral) cavity** (buck′-ul [or′-ul] kav′-i-tee): mouth cavity bounded by the inner surface of the cheek

**bunion** (bun′-yun): swelling of bursa of foot

**bursa** (bur′-suh): small sac interposed between parts that move on one another

**bursitis** (bur-sigh′-tis): inflammation of a bursa

## C

**calcaneus** (kal-kay′-nee-us): heel bone

**calciferol** (kal-sif′-ur-ol): vitamin D$_2$

**calcify** (kal′-si-fī): to deposit mineral salts

**calcitonin** (kal″-si-to′-nin): hormone secreted by thyroid gland which controls calcium ion concentration in body by preventing excessive calcium build up in blood

**calculus** (kal′-kew-lus): stone-like formation in any part of the body, usually composed of mineral salts

**callus** (kal′-us): area of hardened and thickened skin

**calyx** (kay′-liks): cup-shaped part of the renal pelvis

**canine** (kay′-nine): sharp teeth of mammals, between incisors and premolars

**capillary** (kap′-i-lair-ee): microscopic blood vessel which connects arterioles with venules

**carboxyhemoglobin** (kahr-bock″-see-hee′-mo-globin): compound of carbon monoxide and hemoglobin formed when carbon monoxide is present in blood

**carcinoma** (kahr″-si-no′-muh): a malignant tumor

**cardiac** (kahr′-dee-ack): relating to the heart

**cardiac arrest** (kahr′-dee-ack uh-rest′): syndrome resulting from failure of heart as a pump

**cardiac arrhythmia** (kahr′-dee-ack a-rith′-mee-uh): any change or abnormality in the normal heart rhythm or beat

**cardiac muscle** (kahr′-dee-ack mus′-ul): muscle of the heart

**cardiac sphincter** (kahr′-dee-ack sfink′-tur): circular muscle fibers around cardiac end of esophagus

**cardiopulmonary resuscitation** (kahr″-dee-o-puhl′-muh-nair-ee ree-sus″-i-tay′-shun): prevention of asphyxial death by artificial respiration

**caries** (kair′-eez): decay of a tooth or bone

**carotid** (ka-rot′-id): arteries which supply blood to the neck and head

**carpal** (kahr′-pul): bones of the wrist

**cartilage** (kahr′-ti-lidj): white, semi-opaque, nonvascular connective tissue

**casein** (kay′-see-in): protein obtained from milk

**catalyst** (kat′-uh-list): chemical substance which alters a chemical process but does not enter into the process

**cataract** (kat′-uh-rakt): condition in which the eye lens becomes opaque

**cecum** (see′-kum): pouch at the proximal end of the large intestine

**cerumen** (see-roo′-men): ear wax

**cervix** (sur′-vicks): neck; usually the rounded, conical protrusion of the uterus into the vagina

**cholecystectomy** (kol″-e-sis-tek′-tuh-mee): removal of the gallbladder

**cholesterol** (ko-les′-tur-ol): a sterol normally synthesized in the liver and also ingested in egg yolks, animal fats and tissues

**chromosome** (kro′-muh-sohm): nuclear material which determines hereditary characteristics

**chyme** (kime): food which has undergone gastric digestion

**cilia** (sil-ee-uh): tiny lashlike processes of protoplasm

**cochlea** (kock′-lee-uh): spiral cavity of the internal ear containing the organ of Corti

**congenital** (kun-jen′-i-tul): present at birth

**coronary** (kor′-o-nerr″-ee): referring to the blood vessels of the heart

**corpus** (kor′-pus): body

**cortex** (kor′tecks): outer part of an internal organ

**costal** (kos′-tul): pertaining to the ribs

**cretinism** (kree′-tin-ism): congenital and chronic condition due to the lack of thyroid hormone

**cutaneous** (kew-tay′-nee-us): pertaining to the skin

**cyanosis** (si″-uh-no′-sis): bluish color of the skin due to insufficient oxygen in the blood

**cytoplasm** (si′-to-plazm): protoplasm of the cell body, excluding the nucleus

## D

**deciduous teeth** (de-sid′-yoo-us teeth): temporary teeth usually lost by six years of age

**defecation** (def″-e-kay′-shun): elimination of waste material from the rectum

**deglutition** (dee″-gloo-tish′-un): act of swallowing

**deltoid** (del′-toyd): triangular-shaped muscle which covers the shoulder prominence and is used for intramuscular injections in adults

**dendrite** (den′-drīte): nerve cell process that carries nervous impulses toward the cell body

**dentin** (den′-tin): main part of the tooth located under the enamel

**dentition** (den-tish′-un): number, shape, and arrangement of teeth

**deoxygenate** (dee-ock″-si-je-nate): process of removing oxygen from a compound

**dermis** (dur′-mis): true skin; lying immediately beneath the epidermis

**dextrose** (decks′-troce): glucose, a monosaccharide which may accumulate in the urine

**diapedesis** (dye″-uh-pe′dee′sis): passage of blood cells through unruptured vessel walls into tissues

**diaphragm** (dye′-uh-fram): muscular partition between the thorax and the abdomen

**diastole** (dye-as′-tuh-lee): dilation state of the heart; the rest between systoles

**digestion** (di-jes′-chun): change of foods into compounds that can be assimilated

**dislocation** (dis″-lo-kay′-shun): displacement of one or more bones of a joint or of any organ from original position

**distal** (dis′-tul): farthest from point of origin of a structure; opposite of proximal

**dorsal** (dor′-sul): pertaining to the back

**dropsy** (drop′-see): accumulation of serous fluid in a body cavity; edema

**duodenum** (dew″-o-dee′-num): first part of small intestine, beginning at pylorus

**dura mater** (dew′-ruh may′-tur): fibrous membrane forming outermost covering of brain and spinal cord

**dysmenorrhea** (dis-men″-o-ree′-uh): difficult or painful menstruation

**dyspnea** (disp-nee′-uh): labored breathing or difficult breathing

## E

**ectopic** (eck-top′-ick): in an abnormal position; said of an extrauterine pregnancy or cardiac beats

**ectoplasm** (eck′-to-plazm): outer dense layer of cytoplasm of cell or unicellular organism

**edema** (e-dee′-muh): excessive fluid in tissues

**elastic** (e-las′-tick): capable of returning to original form after being compressed or stretched

**elastin** (e-las′-tin): protein base of yellow elastic tissue

**electrolytes** (e-leck′-tro-lights): electrically charged particles which help determine fluid and acid-base balance

**embolism** (em′-bo-lizm): obstruction of a blood vessel by a circulating blood clot, fat globule, air bubble, or piece of tissue

**embryo** (em′-bree-o): the human young up to the first three months after conception; the young of any organism in early development stage

**emesis** (em′-e-sis): vomitus

**emphysema** (em′-fi-see′-muh): lung disorder in which inspired air becomes trapped and is difficult to expire

**empyema** (em″-pye-ee′-muh): pus in a cavity

**endocardium** (en″-do-kahr′-dee-um): membrane lining interior of heart

**endocrine** (en′-do-krin): pertaining to a gland which secretes into the blood or tissue fluid instead of into a duct

**endometrium** (en″-do-mee′-tree-um): mucous membrane lining uterus

**endoplasm** (en′-do-plazm): inner cytoplasm of cell or unicellular organism

**enzyme** (en′-zime): organic catalyst that initiates and accelerates a chemical reaction

**eosinophil** (ee″-o-sin′-uh-fil): white blood cell or bone marrow whose granules stain red with eosin or other acid dyes

**epidermis** (ep″-i-dur′-mis): outermost layer of skin

**epididymitis** (ep″-i-did″-i-migh′-tis): inflammation of epididymis

**epiglottis** (ep″-i-glot′-is): elastic cartilage covered by mucous membrane forming upper part of larynx which guards glottis during swallowing

**epinephrine** (ep″-i-nef′-rin): adrenalin; secretion of the adrenal medulla, which prepares the body for energetic action

**erythrocyte** (e-rith′-ro-sight): red blood cell

**erythropoiesis** (e-rith″-ro-poy-ee′-sis): formation or development of red blood cells

**exophthalmos** (eck″-sof-thal′-mus): abnormal protrusion of the eyes

**expiration** (eck″-spi-ray′-shun): act of breathing forth or expelling air from lungs

**extensor** (eck-sten′-sur): muscle which extends or stretches a limb or part

## F

**fallopian tube** (fa-lo′-pee-un tewb): uterine tube or oviduct which carries egg from ovary to uterus

**fascia** (fash′uh): band or sheet of fibrous membranes covering or binding and supporting muscles

**femur** (fee′-mur): thighbone

**fetus** (fee′-tus): the human young from birth until the third month of the intrauterine period

**fibrin** (fī'-brin): an insoluble protein necessary for the clotting of blood

**fibrinogen** (fī-brin'-o-jen): a protein which is converted into fibrin by the action of thrombin

**fibroma** (fī-bro'-muh): benign tumor made up mainly of fibrous connective tissue

**fibula** (fib'-yoo-luh): slender bone at outer edge of lower leg

**fontanel** (fon"-tuh-nel'): unossified areas in the infant skull; soft spot

**foramen** (fo-ray'-men): an opening in a bone

**fracture** (frack'-chur): a break in a bone

**fundus** (fun'-dus): part farthest from opening of an organ

## G

**ganglion** (gang'-glee-un): a mass of nerve cell bodies outside the central nervous system

**gangrene** (gang'-green): necrosis of a part due to failure of blood supply

**gastric** (gas'-trick): pertaining to the stomach

**gastric glands** (gas'-trick glans): glands lining stomach

**gene** (jeen): part of the chromosome that transmits a specific hereditary trait

**genitals** (jen'-i-tuls): reproductive organs, also called genitalia

**gestation** (jes-tay'-shun): development period of the human young from conception to birth

**glenoid fossa** (glee'-noid fos'-uh): articular surface on scapula for articulation with head of humerus

**glomerulus** (glom-err'-yoo-lus): compact cluster of capillaries in the nephron of the kidney

**glucose** (gloo'-koce): a monosaccharide or simple sugar; the principal blood sugar

**gluteal** (gloo-tee'-ul): pertaining to the area near the buttocks

**glycerin or glycerol** (glis'-ur-in or glis'-ur-ole): product of fat digestion

**glycogen** (glye'-kuh-jin): polysaccharide formed and stored largely in the liver; can be converted into glucose when needed

**gonads** (go'nads): sex glands (ovaries or testes)

**graafian follicle** (graf'-ee-un fol'-i-kul): mature ovarian follicle

**greenstick fracture** (green-stick frack'-chur): incomplete fracture of long bone; seen in children; bone is bent but splintered only on convex side

## H

**hemiplegia** (hem"-i-plee'-jee-uh): paralysis of one side of the body

**hemocytoblast** (hee"-mo-sigh'-to-blast): cell considered by some to be primitive stem cell, giving rise to all blood cells

**hemoglobin** (hee'-muh-glo"-bin): oxygen-carrying pigment of the blood

**hemophilia** (hee"-mo-fill'-ee-uh): sex-linked, hereditary bleeding disorder occurring only in males but transmitted by females; characterized by a prolonged clotting time and abnormal bleeding

**hemorrhoids** (hem'-uh-roydz): enlarged and varicose condition of the veins in the lower part of the anus or rectum and the tissues of the anus

**heparin** (hep'-uh-rin): substance obtained from the liver, which slows blood clotting

**hepatic vein** (he-pat'-ick vain): vein which drains blood from liver into inferior vena cava

**hernia** (hur'-nee-uh): protrusion of a loop of an organ through abnormal opening

**histology** (his-tol'-uh-jee): microscopic study of living tissues

**hormone** (hor'mone): chemical secretion, usually from an endocrine gland

**humerus** (hew'-mur-us): upper arm bone

**hyoid bone** (high'-oyd bone): bone between root of the tongue and larynx, supporting tongue and giving attachment to several muscles

**hyperopia** (high"-pur-o'-pee-uh): farsightedness

**hypertension** (high"-pur-ten'-shun): abnormally high blood pressure

**hypertrophy** (high-pur'-truh-fee): enlargement of a part due to increase in size of its already existing cells

**hypotension** (high"-po-ten'-shun): reduced or abnormally low tension, synonymous with low blood pressure

**hysterectomy** (his"-tur-eck'-tuh-mee): partial or total surgical removal of the uterus

## I

**ilium** (il'-ee-um): upper broad portion of the hipbone

**incisor** (in-sigh'-zur): cutting tooth; one of four front teeth of either jaw

**incus** (ing'-kus): the middle ear bone, also called the anvil

**ingestion** (in-jes′-chun): act of taking substances, especially food, into body

**inguinal** (ing′-gwi-nul): pertaining to the groin

**inhalation** (in″-huh-lay′-shun): taking air into the lungs

**innominate bone** (i-nom′-i-nut bone): hipbone

**inspiration** (in″-spi-ray′-shun): drawing in of air; inhalation

**insulin** (in′-suh-lin): hormone secreted by the pancreas; regulates the rate of carbohydrate usage

**integument** (in-teg′-goo-munt): covering, especially the skin

**intercostal muscles** (in″-tur-kos′-tul mus′-ul): muscles found between adjacent ribs

**involuntary** (in-vol′-un-terr-ee): opposite of voluntary, not within the control of will

**involution** (in″-vo-lew′-shun): return of an organ to its normal size after enlargement; also the regressive changes due to aging

**ion** (eye′-on): an electrically charged atom

**iris** (eye′-ris): colored, circular smooth muscle surrounding the pupil and controlling the diameter of pupil

**irritability** (irr″-i-tuh-bil′-i-tee): ability to react to a stimulus; excitability

**ischium** (is′-kee-um): lower part of hipbone

**isotonic** (eye″-so-ton-ick): the same tension or pressure

# K

**keratectomy** (kerr″-uh-teck′-tuh-mee): excision of part of the cornea

**keratin** (kerr′-uh-tin): chemical belonging to albuminoid or scleroprotein group found in horny tissue, hair, nails, feathers; insoluble in protein solvents, and containing a high percentage of sulfur

**kilogram** (kil′-uh-gram): 1000 grams or approximately 2.2 pounds

**kinesthesia** (kin″-es-thee′-shuh): ability to perceive the direction or weight of muscular movement

**kinetic** (ki-ne′-tick): pertaining to motion

# L

**labia** (lay′-bee-uh): lips

**lacrimal** (lack′-ri-mul): pertaining to tears

**lactation** (lack′-tay′-shun): secretion of milk from the breasts

**lactose** (lack′-toce): milk sugar; a disaccharide used in infant formulas

**larynx** (lar′-inks): voicebox, found between trachea and base of tongue, containing the vocal cords

**lateral** (lat′-ur-ul): toward the side

**laxative** (lack′-suh-tiv): chemical substance that relieves constipation; a mild purgative

**lens** (lenz): crystal or glass for refraction of light rays

**leukocyte** (lew′-ko-sight): white blood cell

**leukorrhea** (lew″-ko-ree′-uh): whitish, mucopurulent discharge from vagina

**ligament** (lig′-uh-munt): a band of fibrous tissue connecting bones or supporting organs

**lipid** (lip′-id): fatty compound

**locomotion** (lo″-kuh-mo′-shun): act of moving from place to place

**lordosis** (lor-do′-sis): forward curvature of lumbar region of spine

**lumbar** (lum′-bahr): pertaining to the loins; region between the posterior thorax and sacrum

**lumbar vertebrae** (lum′-bur vur′-te-bree): five vertebrae associated with lower part of back

**lumen** (lew′-min): passageway or opening to a tubular structure such as a blood vessel

**lymph** (limf): watery fluid in the lymphatic vessels

**lymphatic** (lim-fat′-ick): vessel carrying lymph

**lymphatic system** (lim-fat′-ick sis′-tum): system of vessels and nodes supplemental to blood circulatory system, carrying lymph

**lymphocyte** (lim′-fo-sight): a type of white blood cell

**lysosome** (lye″-so-sohm): cytoplasmic organelle present in many kinds of cells, especially in liver and kidney cells containing digestive enzymes

# M

**malleus** (mal′-ee-us): largest of three middle ear bones; also called the hammer

**maltose** (mawl′-toce): disaccharide formed by the hydrolysis of starch

**mammary** (mam′-ur-ree): pertaining to the breast

**mandible** (man′-di-bul): lower jawbone

**manubrium** (ma-new′-bree-um): (1) handle-like process; (2) upper part of the sternum (breastbone)

**mastectomy** (mas-teck'-tuh-mee): amputation of breast

**mastication** (mas"-ti-kay'-shun): the process of chewing

**maturation** (match"-oo-ray'-shun): process of coming to full development

**meatus** (mee-ay'-tus): passageway or opening

**medial** (mee'dee-ul): toward midline of body

**mediastinum** (mee"-dee-as-tye'-num): intrapleural space; separating the sternum in front and the vertebral column behind

**medulla** (ma-dul'-uh): inner portion of an organ

**melanism** (mel'-uh-nizm): abnormal deposition of dark pigment (melanin) in organs, in tissues, or in skin

**membrane** (mem'-brane): a thin layer of tissue which covers a surface or divides an organ

**menarche** (me-nahr'-kee): time when menstruation begins

**menopause** (men'-o-pawz): physiologic termination of menstruation, generally between forty-fifth and fiftieth years

**menstruation** (men"-stroo-ay'-shun): monthly discharge of blood from the uterus; also called menses

**metabolism** (me-tab'-o-lizm): sum total of processes of digestion, absorption, and the resulting release of energy

**metacarpus** (met"-uh-kahr'-pus): part of the hand between the wrist and the fingers

**metastasis** (me-tas'-tuh-sis): transfer of disease from an original site to a distant one by the transmittal of causative agents or cells through the circulatory system or lymph vessels

**metatarsus** (met"-uh-tahr'-sus): part of the foot between the tarsal bones and the toes

**microbe** (mi'-krobe): microscopic organisms, especially bacterium

**microcephalus** (migh"-kro-sef'-uh-lus): individual with an unusually small head

**micturition** (mich"-tew-rish'-un): voiding, or urinating

**miotic** (mi-ot'-ick): causing contraction of pupil

**mitosis** (mi-to'-sis): cell division into two new cells, each possessing a complete set of chromosomes

**mixed nerve** (mikst nurv): nerve composed of both afferent (sensory) fibers and efferent (motor) fibers

**monocyte** (mon'-o-sight): large mononuclear leukocyte with deeply indented nucleus, slate-gray cytoplasm, and fine bluish granulations

**monosaccharide** (mon"-o-sack'-uh-ride): simple sugar; glucose

**mucin** (mew'-sin): mixture of glycoproteins forming basis of mucus

**mucosa** (mew-ko'-suh): mucous membrane

**myelin** (migh'-e-lin): a lipoid substance found in the sheath around nerve fibers

**myocarditis** (mi"-o-kahr-dye'-tis): inflammation of muscular tissue of heart

**myocardium** (mi'-o-kahr'-dee-um): muscle of the heart

**myoglobin** (mi'-o-glo"-bin): form of hemoglobin occurring in muscle fibers

**myometrium** (mi"-o-mee'-tree-um): uterine muscular structure

**myopia** (mi-o'-pee-uh): nearsightedness

## N

**nares** (nair'-eez): pertaining to the nostrils

**nasal cavity** (nay'-zul kav'-i-tee): one of the pair of cavities between anterior nares and nasopharynx

**nasal septum** (nay'-zul sep'-tum): partition between the two nasal cavities

**nephron** (nef'-ron): unit of structure and function of kidney, consisting of glomerular capsule, glomerulus, and attached kidney tubules

**neuron** (new'-ron): nerve cell, including its processes

**nongravid** (non"-grav'-id): not pregnant

**nonpathogenic** (non"-path"-o-jen'-ick): incapable of producing disease

**nucleolus** (new-klee'-uh-lus): small spherical structure within cell nucleus

**nucleus** (new'-klee-us): core or center of a cell containing large quantities of DNA

## O

**occiput** (ock'-si-put): pertaining to the back of the head

**olecranon process** (o-leck'-ruh-non pro'-sess): large projection at upper extremity of ulna

**olfactory** (ol-fack'-tur-ee): pertaining to the sense of smell

**oogenesis** (o"-o-jen'-e-sis): process of origin, growth, and formation of ovum in ovary during preparation for fertilization

**ophthalmic** (off-thal'-mick): referring to the eyes

**orchitis** (or-ki'-tis): inflammation of testis

**organelle** (or-guh-nel'): microscopic specialized structure, or part of cell having a special function or capacity

**oropharynx** (or"-o-far'-inks): oral pharynx, found below level of lower border of soft palate and above larynx

**osmosis** (oz-mo'-sis): passage of fluid through a membrane

**osmotic pressure** (oz-mot'-ick presh'-ur): pressure developed when two solutions of different concentrations of the solute are separated by a membrane permeable only to the solvent

**ossa carpi** (os-sa kahr'-pye): the eight bones of the wrist

**osseous** (os'-ee-us): bony; composed of or resembling bone

**ossicle** (os'-i-kul): a small bone; usually refers to the three small bones of the middle ear

**osteoarthritis** (os"-tee-o-ahr-thrigh'-tis): degenerative joint disease

**osteoblast** (os'-tee-o-blast): cells involved in formation of bony tissue

**osteoclast** (os'-tee-o-klast): cells involved in resorption of bony tissue

**oxygenate** (ock'-si-ji'-nate): to saturate a substance with oxygen, either by chemical combination or by mixture

**oxyhemoglobin** (ock"-si-hee'-muh-glo"-bin): hemoglobin combined with oxygen

# P

**palate** (pal'-ut): roof of the mouth

**Pap or Papanicolaou smear**: cytological, diagnostic cancer technique studying exfoliated cells, especially those from the vagina

**papilla** (pa-pil'-uh): small, nipple-shaped elevations

**paralysis** (puh-ral'-i-sis): loss of power of motion or sensation

**parathyroid gland** (par"-uh-thi'-royd gland): one of several (usually four) small endocrine glands embedded in the thyroid gland, secreting a hormone which regulates blood calcium levels (parathormone)

**paresthesia** (par"-es-theezh'-uh): perverted sensation of tingling, crawling, or burning of skin

**parotid gland** (pa-rot'-id): largest of the salivary glands

**patella** (pa-tel-uh): kneecap

**pectoral** (peck'-tuh-rul): pertaining to the chest

**pelvis** (pel'-vis): any basin-shaped structure or cavity

**pericardium** (perr"-i-kahr'-dee-um): closed membranous sac surrounding heart

**peripheral** (pe-rif'-e-rul): outside surface, or the area away from the center

**peristalsis** (perr"-i-stal'-sis): progressive wave of contraction in tubular structures provided with longitudinal and transverse muscular fibers, as in esophagus, stomach, small and large intestines

**pH**: hydrogen ion concentration of solution or air mixture; potential of hydrogen

**phagocyte** (fag'-o-sight): cell having property of engulfing and digesting foreign particles or cells harmful to body

**phagocytosis** (fag"-o-si-to'-sis): ingestion of foreign or other particles by certain cells

**phalanges** (fa-lan'-jeez): bones of fingers and toes

**pharynx** (far'-inks): musculomembranous tube located behind nose, mouth, and larynx, extending from base of skull to point opposite sixth cervical vertebra

**phlebitis** (fle-bye'-tis): inflammation of a vein, with or without infection and thrombus formation

**physiology** (fiz"-ee-ol'-uh-jee): science that studies functions of living organisms and their parts

**physiotherapy** (fiz"-ee-o-therr'-uh-pee): physical medicine

**pia mater** (pee'uh may-tur): vascular innermost covering of brain and spinal cord

**pigment** (pig'-munt): (1) dye or coloring matter; (2) organic coloring matter of body

**plasma** (plaz'-muh): liquid part of blood containing corpuscles

**pleura** (ploor'-uh): serous membrane enclosing lung and lining internal surface of thoracic cavity

**pleurisy** (ploor'-i-see): inflammation of pleura

**plexus** (pleck'-sus): network of interlacing nerves, blood vessels, or lymphatics

**polysaccharide** (pol"-ee-sack'-uh-ride): a complex sugar

**popliteal** (pop-lit'-ee-ul): area behind knee

**posterior** (pos-teer'-ee-ur): located behind or at the back; opposite to anterior

**presbyopia** (prez"-bee-o'-pee-uh): farsightedness of advanced age due to loss of elasticity in lens of eye

**progesterone** (pro-jes'-tur-ohn): steroid hormone secreted by ovary from corpus luteum to help maintain pregnancy

**pronation** (pro-nay'-shun): (1) condition of being prone, (2) turning of palm of hand downward

**prostatectomy** (pros"-tuh-teck'-tuh-mee): surgical removal of all or part of prostate

**protoplasm** (pro'-tuh-plazm): living colloid material of the cell; contains proteins, lipids, inorganic salts and carbohydrates

**proximal** (prock'-si-mul): located nearest the center of the body; point of attachment of a structure

**puberty** (pew'-bur-tee): age when reproductive organs become functional

**pubis** (pew'-bis): pubic bone, portion of hipbone forming front of pelvis

**pulmones** (pul'-mones): plural of lung (pulmo)

**pupil** (pew'-pil): opening in iris of eye for passage of light

**pylorus** (pye-lo'-rus): circular opening of stomach into duodenum

## Q

**quadripedal** (kwah'-dri-pe-dul): four-footed stance

## R

**receptor** (re-sep'-tur): sensory nerve that receives a stimulus and transmits it to the CNS

**red muscle** (red mus'ul): muscle that appears red in fresh state, due to presence of muscle hemoglobin

**reflex** (ree'-flecks): involuntary action; automatic response

**reflex arc** (ree'-flecks ahrk): pathway travelled by an impulse during reflex action, going from receptor to effector

**reflux** (ree'-flucks): return flow

**renal** (ree'-nul): pertaining to the kidney

**rennin** (ren'-in): milk-coagulating enzyme found in gastric juice of ruminating animals, not present in the human stomach

**retroperitoneal** (ret"-ro-perr"-i-to-nee'-ul): located behind the peritoneum

**ribosome** (rye'-bo-sohm): submicroscopic particle attached to endoplasmic reticulum, site of protein synthesis in cytoplasm of cell

**rugae** (roo'-jee): wrinkles or folds

## S

**sacroiliac joint** (say"-kro-il'-ee-ack joint): joint between sacrum and ilium

**sagittal** (sadj'-i-tul): longitudinal; shaped like an arrow

**sartorius** (sahr-to'-ree-us): thigh muscle

**scapula** (skap'-yoo-luh): large, flat, triangular bone forming back of shoulder

**sclera** (skleer'-uh): tough, white covering; part of external coat of eye

**scoliosis** (sko"-lee-o'-sis): lateral curvature of the spine

**scrotum** (skro'-tum): pouch that contains the testicles

**sebaceous gland** (se-bay'-shus gland): gland that secretes sebum, a fatty material

**sebum** (see'-bum): secretion of sebaceous glands that lubricate the skin

**sella turcica** (sel'-uh tun'-si-kuh): saddle-shaped depression in sphenoid bone

**semen** (see'-mun): male reproductive fluid containing sperm

**semilunar** (sem"-ee-lew'-nur): half-moon shaped valve of aorta and pulmonary artery

**senescence** (se-nes'-unce): old age; senility

**septum** (sep'-tum): partition; dividing wall between two spaces or cavities, such as the septum between left and right side of heart

**serum** (seer'-um): clear, pale yellow fluid that separates from a clot of blood; plasma that contains no fibrinogen

**sigmoid** (sig'-moid): shaped like the letter S; distal, S-shaped part of colon

**sinoatrial node** (sigh"-no-ay'-tree-ul node): dense network of Purkinje fibers of conduction system at junction of superior vena cava and right atrium

**sinus** (sigh'-nus): recessed cavity or hollow space

**skeletal muscle** (skel'-e-tul mus'-ul): muscle attached to a bone or bones of skeleton and concerned in body movements; also known as voluntary or striated muscle

**solute** (sol'-yoot): dissolved substance in a solution

**sphincter** (sfink'-tur): circular muscle, such as the anus

spina bifida (spye'-nuh bye'-fi-duh): congenital defect in closure of spinal canal with hernial protrusion of meninges of spinal cord

sprain (sprain): wrenching of a joint, producing a stretching or laceration of ligaments

stapes (stay'-peez): stirrup-shaped bone in middle ear

sternum (stur'-num): flat, narrow bone in median line in front of chest, composed of three parts: manubrium, body, and xiphoid process

stethoscope (steth'-uh-skope): instrument used for detection and study of sounds arising within body

subluxation (sub''-luck-say'-shun): incomplete dislocation

sudoriferous (sue''-dur-if'-ur-us): producing perspiration

superior (sue-peer'-ee-ur): in anatomy, higher; denoting upper of two parts, toward vertex

supination (sue''-pi-nay'-shun): turning of palm of hand upward; condition of being supine (lying on back)

surfactant (sur-fack'-tunt): surface-active agent

synapse (sin'-ops): space between adjacent neurons through which an impulse is transmitted

syncytium (sin-sish'-ee-um): mass of cytoplasm with numerous nuclei

synovia (si-no'-vee-uh): viscid fluid present in joint cavities

synthesis (sinth'-e-sis): in chemistry, processes and operations necessary to build up a compound; in general, a reaction, or series of reactions, in which a complex compound is obtained from elements or simple compounds

systole (sis'-tuh-lee): contraction of ventricles, forcing blood into aorta and pulmonary artery

# T

talus (tay'-lus): ankle bone that articulates with bones of leg

tarsus (tahr'-sus): instep

tendon (ten'-dun): cord of fibrous connective tissue that attaches a muscle to a bone or other structure

tetanus (tet'-uh-nus): infectious disease, usually fatal, characterized by spasm of voluntary muscles and convulsions caused by toxin from tetanus bacillus, **clostridium tetani**

thoracocentesis (thor''-ruh-ko-sen-tee'-sis): aspiration of chest cavity for removal of fluid, usually for empyema

thorax (tho'-racks): chest; portion of trunk above diaphragm and below neck

thrombin (throm'-bin): enzyme found in blood produced from an inactive precursor, prothrombin, inducing clotting by converting fibrinogen to fibrin

thrombosis (throm-bo'-sis): formation of a clot in a blood vessel

thyroxine (thigh-rock'-seen): hormone secreted by thyroid gland or prepared synthetically; contains about 64% iodine

tibia (tib'-ee-uh): larger, inner bone of the leg, below the knee

transverse (trans-vurce'): crosswise; at right angles to longitudinal axis of body

triceps (trye'-seps): (1) three-headed; (2) muscle having three heads, as triceps brachii

turbinate (tur'-bin-ut): shaped like a spiral; the three bones situated on the lateral side of the nasal cavity

tympanum (tim'-puh-num): drum; middle ear closed externally by the ear drum

# U

ulna (ul'-nuh): bone on inner forearm

umbilicus (um-bil'-i-kus): navel

unicellular (yoo''-ni-sel'-yoo-lur): composed of one cell

unilateral (yoo''-ni-lat'-ur-ul): pertaining to, or affecting, one side

universal donor (yoo''-ni-vur'-sul re-sip'-ee-unt): individual belonging to AB blood group

uvula (yoo'-vew-luh): projection hanging from soft palate

# V

vacuole (vack'-yoo-ole): (1) clear space in cell; (2) cavity bound by a single membrane; usually a storage area for fat, glycogen, secretions, liquid, or debris

vagina (va-jie'-nuh) sheathlike structure; tube in females, extending from the uterus to the vulva

**valve** (valv): structure which permits flow of a fluid in only one direction

**varicose veins** (var'-i-koce vains): veins that have become abnormally dilated and tortuous, due to interference with venous drainage or weakness of their walls

**vein** (vane): vessel which carries blood toward the heart

**ventral** (ven'-trul): front or anterior; opposite of posterior or dorsal

**ventricle** (ven'-trik-ul): small cavity or chamber, as in heart or brain

**venule** (ven'-yool): small vein

**vermiform appendix** (vur'-mi-form a-pen'-dicks): small, blind gut projecting from cecum

**villi** (vil'-eye): hairlike projections, as in intestinal mucous membrane

**viscera** (vis'-er-a): internal organs

**vitreous humor** (vit'-ree-us hew'-mur): transparent, gelatin-like substance filling greater part of eyeball

**voluntary** (vol'-un-terr"-ee): under willful control

## W

**white muscle** (white mus'-ul): skeletal muscle that appears paler in fresh state than red muscle

**wisdom tooth** (wiz'-dum tooth): third molar tooth in adult mouth

## Z

**zona pellucida** (so'-nuh pe-lew'-si-duh): thick, solid, elastic envelope of ovum

**zygote** (zye'-gote): organism produced by union of two gametes

# Index

Abdominal cavity, 4, 5, 164, 174, 183, 184
Abdominal hernia, 90
Abdominopelvic cavity, 4, 5
Abduction, 47, 48, 83
Absolute zero, 13
Abnormal muscular conditions, 90, 91
Accessory organs, digestion, 165
Acetabulum, 63
Acne vulgaris, 212-213
Acromegaly, 258, 259
ACTH, 241, 242, 250
Active transport, 16, 17
Acute bacterial meningitis, 289
Acute kidney failure, 211
Acute nephritis, 212
Acute rheumatic heart disease, 132
Addison's disease, 250, 259
Adduction, 47, 48, 83
Adenitis, 129
Adenoid, 144
Adenosine triphosphate, 16, 17
ADH, 201, 242, 272
Adipose, 30
Adrenal glands, 238, 239, 250-252
  disorders, 259
Adrenalin, 251
Adrenocorticotrophic hormone, 241, 242, 250
Afferent nerve, 275
Agglutinin, 121, 122
Agglutinogen, 121, 122
Agranulocyte, 119
Albinism, 207
Albumin in urine, test for, 202, 203
Aldosterone, 250
Alimentary canal, 164, 165, 180
Alveolar sacs, 146
Alveoli, 140, 141, 143, 146, 150
Anabolism, 3
Anal sphincter, 78, 184, 185
Androgens, 251
Anemia, 134
Aneurysm, 134
Angina pectoris, 133
Ankylosis, 69
Anterior lobe, 241, 242
Anticoagulant, 120
Anticonvulsant, 289
Antidiuretic hormone, 201, 242, 272
Antiprothrombin, 120
Antithromboplastin, 120
Anuria, 211
Anvil, 284, 285
Aorta, 95, 96, 105, 109
Aortic semilunar valve, 100, 109
Apex of the lung, 146

Aponeurosis, 31
Appendicitis, 183
Appendicular skeleton, 59-64
Appendix, vermiform, 183, 184
Aqueous humor, 283
Arachnoid, 267
Arbovirus, 288
Areola, 229
Arm, 56, 59, 61
Arrhythmia, 132
Arteries, 104, 105, 111, 112, 114, 115, 118, 134
Arterioles, 104, 105, 112, 113
Arteriosclerosis, 111, 134
Arthritis, 69, 70
Ascites, 133
Assimilation, 3
Asthma, 158
Atelectasis, 158
Athlete's foot, 213
Atlas vertebra, 58, 60
ATP, 16, 17
Atrium, 99, 100, 104-106, 108
Atrial fibrillation, 132
Atrioventricular node, 100
Atrioventricular valve, 99
Atrophy, 90
Auricle, 99
Autonomic nervous system, 265, 275-277
Autosome, 220
A-V node, 100
Axial skeleton, 56
Axilla, 129
Axis vertebra, 58, 60
Axon, 265

Ball-and-socket joints, 47, 63
Basal metabolism rate, 256, 257
Base, 95
Basophil, 119
Belly, 82
Benign, 190, 233
Beta cells, 253, 254
Bicarbonate ion ($HCO_3$), 141
Biceps femoris muscle, 86, 88
Biceps muscles, 82, 86, 87
Bicuspid (mitral) valve, heart, 99, 100, 105, 109, 133
Bicuspids, teeth, 170, 171
Bifurcation, 104
Bile, 180, 181, 189
Biopsy, 233
Bipedal, 58
Bladder, 198, 202, 211, 212

Blood
  composition, 94, 95, 116-120
  disorders, 134, 135
  general circulation, 104-106
  lymphatic system, 127-129
  pulmonary circulation, 96, 108, 109
Blood circulatory system, 94-96, 98-100, 104-106, 108, 109, 111-123
Blood clotting, 119-121
Blood norms, 122, 123
Blood plasma, 94, 95, 116, 117, 121, 122, 127
Blood platelets, 94, 116, 119, 120, 122
Blood pressure, 114
  disorders, 135
Blood type, 121
Blood vessels, 94, 95, 111-115, 120, 128
  disorders, 134
Body, whole, 2-8, 9-18, 27-33, 35-37
Body balance, 3
Body framework, 44-48, 51-53, 56-65
  bone structure, 46, 47, 51-53
  injuries and disorders, 68-71
  skeleton, 56-65
Body functions, coordination, 264-273, 275-278, 281-285
  central nervous system, 267-273
  disorders of nervous system, 287-290
  peripheral and autonomic nervous systems, 275-278
  sense organs — eye and ear, 281-285
Body functions, regulators, 238-254
  adrenal glands and gonads, 250, 251
  disorders, 256-260
  endocrine system, 238, 239
  pancreas, 238, 239, 253, 254
  pituitary gland, 238, 239, 241, 242
  thyroid and parathyroid gland, 238, 239, 242, 245-247
Body movement, 76-78, 81-83, 86-88
  abnormal muscular conditions, 90, 91
  attachment of muscles, 81-83
  muscular system, 76-78
  skeletal muscles, 86-88
Body systems, 35-37
Boils, 214
Bolus, 165, 173
Bones, skeletal system, 44-48, 51-53, 56-65
  injuries and disorders, 68-71
  skeleton, 56-65
  structure and formation, 46, 47, 51-53

307

tissue, 31
Bowel, 185, 186, 190
Bowman's capsule, 199, 200
Brachial, 112
Brachiocephalic artery, 105
Brain, 264, 265, 267-272, 276
Brainstem, 273
Breasts, 225, 226, 229, 233
    mastectomy, 233
Breathing processes, 140-154
    disorders, 156-158
    mechanics of breathing, 150-154
    organs and structures, 143-147
Bright's disease, 212
Broken bones, 68, 69
Bronchi, 145, 146, 157, 158
Bronchiectasis, 158
Bronchioles, 143, 146
Bronchitis, 157
Buccal, 6, 165
Bulbourethral glands, 229, 230
Bundle of His, 100
Bunion, 70
Bursa, 46, 70
Bursitis, 70

Calcaneous, 64
Calciferol (vitamin D), 70, 247, 248
Calcify, 31, 70
Calcitonin, 246, 247, 248
Calculus, 212
Callus, 207
Calyx, 199
Cancer,
    digestive tract, 190, 191
    larynx, 158
    lung, 158
    reproductive system, 129, 233, 234
Canine teeth, 170, 171
Capillaries, 95, 104, 108, 111-113, 118, 127
Carboxyhemoglobin, 118
Carbuncles, 214
Carcinoma. See Cancer
Cardiac arrest, 98, 99
Cardiac arrhythmia, 132
Cardiac cycle, 100
Cardiac muscle, 78, 99, 133
Cardiac sphincter, 174
Cardiopulmonary resuscitation (CPR), 99
Carotid, 105, 112, 115
Carpal bones, 57, 61
Cartilage tissues, 28, 31, 32, 46, 51, 53, 56, 58, 59, 61, 69
Cartilaginous joints, 31, 46, 47, 69
Casein, 177
Catabolism, 3
Cataract, 290
Cathartic, 186
Cecum, 183, 184
Cell membrane, 10, 11, 12-14, 17, 264
Cells, 9-18
Cellular respiration (oxidation), 12, 141
Central nervous system, 265, 267-273, 275, 276
Cerebellum, 270, 272, 273
Cerebral hemorrhage, 288
Cerebral palsy, 288
Cerebrum, 267-272
Cervical vertebra, 58, 59

Cervix, 227, 228, 234
Chief cells, 176
Chemical reactions, 10, 12
Cholecystitis, 190
Chorea, 287
Choroid coat, 281, 283
Chromatin material, 10
Chromosomes, 10, 220-223
Chronic constipation, 186, 190
Chronic nephritis, 212
Chyme, 177
Cilia, 29, 143, 146
Circulatory system, 37, 94-129
    blood, 94-96, 116-123
    blood vessels, 111-115
    disorders, 132-135
    general, 96, 104, 105
    heart, 98-100, 104-105
    introduction to, 94-96
    lymphatic system, 127-129
    pulmonary, 108-109
Cirrhosis, 189
Clavicles, 57, 59, 61
Cleavage, 222
Closed circulatory system, 94, 95
Clotting time, 121
Clubfoot, 70
Coagulation, 120
Coccyx, 58, 59
Cochlea, 284, 285
Cochlear duct, 284
Coitus, 220
Collagen, 30
Colon, 183, 184
Colonic stasis, 186
Comminuted fracture, 68, 69
Common cold, 156
Compact bone, 52
Compliance, 151
Compound fracture, 68
Concussion, 56
Conductivity, 264
Cones, of eye, 284
Congenital heart disease, 132
Congenital malformations, 70
Congestive heart failure, 132
Conjunctivitis, 289
Connecting neurons, 265
Connective tissue, 27, 30-32
Constipation, 186
Contraction, 81
Convolutions, brain, 269
Convulsions, 288
Cooley's anemia, 135
Coordination of body functions, 264-290
    central nervous system, 267-273
    disorders of nervous system, 287-290
    nervous system, 264-265
    peripheral and autonomic nervous systems, 275-278
    sense organs — eye and ear, 281-284
Cornea, 282
Corona radiata, 221
Coronary artery, 105
Coronary circulation, 105
Coronary occlusion, 134
Coronary sinus, 105
Corpus luteum, 226
Cortex, 250, 269
Corticoids, 250, 251

Cowper's glands, 229
Cranial cavity, 4
Cranial nerves, 275
Cranium, 56
Cretinism, 257
Cushing's Syndrome, 259
Cusps, 99
Cyclosis, 10
Cystitis, 103
Cytoplasm, 10, 11

Deciduous teeth, 169
Deglutition, 173
Deltoid muscle, 86, 87
Dendrites, 264
Deoxygenated blood, 108
Deoxyribonucleic acid (DNA) 10, 222
Depressed fracture, 56
Dermis, 205
Descending colon, 184
Diabetes insipidus, 259
Diabetes mellitus, 260
Diaphragm, 4
    muscle, 86
Diaphysis, 52
Diarrhea, 190
Diastolic blood pressure, 114
Diffusion, 14
Digestive system, 164-186
    disorders, 188-191
    introduction to, 164-170
    large intestine, 183-186
    small intestine, 180-181
    stomach, 173-177
Diphtheria, 157
Diseases. See Disorders
Dislocation, 68
Disorders, of
    bones and joints, 68
    circulatory system, 132
    digestive system, 188
    endocrine system, 256
    excretory system, 211
    musculoskeletal system, 90
    nervous system, 287
    reproductive system, 233
    respiratory system, 156
Distal convoluted tubule, 199
DNA, 10, 222
Dorsal, 6
Dorsal cavity, 4
Ductless glands, 238
Duct of Bartholin, 168
Ducts of Rivinus, 168
Duodenum, 174, 180
Dura mater, 267
Dwarfism, 259
Dysmenorrhea, 234

Ear, 284-285
    disorders, 289
Ectopic pregnancy, 227
Ectoplasm, 10
Eczema, 213
Effectors, 278
Efferent nerve, 275
Elastin, 30
Electrolyte, 117
Electron microscope, 4, 12
Elimination of waste materials, 196-214
    disorders, 211-214

excretory system, 196-197
  skin, 205-209
  urinary system, 198-202
Embolism, 135
Embryo, 228, 229
Emphysema, 158
Encephalitis, 288
End organs of Krause, 207
End organs of Ruffini, 207
Endocarditis, 133
Endocardium, 99
Endocervicitis, 234
Endocrine glands, 238, 239
Endocrine system, parts of, 37, 238-239
  adrenal glands and gonads, 250-252
  disorders, 256-260
  pancreas, 253-254
  pituitary gland, 241-242
  thyroid and parathyroid glands, 245-248
Endometrium, 226, 228
Endoplasm, 10
Endoplasmic reticulum, 10, 11-12
Endosteum, 52
Energy, 2, 12
  oxidation process, 83, 140
Enzymes, 164
  pepsin, 176
  small intestine, 180
Eosinophil, 119
Epidermis, 205, 206-207
Epididymis, 229, 230
Epididymitis, 234
Epiglottis, 173
Epilepsy, 289
Epinephrine, 251
Epiphysis, 52
Epithelial tissue, 27, 29
Equilibrium, 284
ERV, 154
Erythrocytes, 53, 118, 123
Erythropoiesis, 118
Estrogen, 226, 251
Ethmoid sinus, 143
Eustachian tube, 284, 285
Excretion, 185
Excretory system, 37, 196-214
  disorders, 211-214
  introduction to, 196
  skin, 205-209
  urinary system, 198-202
Exhalation, 150
Expiration, 150
Expiratory reserve volume, 154
Extension, 47
Extensor muscles, 82, 86, 87-88
External os of cervix, 228
External respiration, 140
Extrinsic muscles, 282
Eye,
  disorders, 289, 290

Facial bones, 56
Fallopian tube, 225, 226, 227
False rib, 59, 61
Familial hemolytic jaundice, 135
Feet, bones, 64
  flatfoot, 90
Female reproductive system, 225-229
Femur, 63
Fertilization, 222-223
Fetus, 229
Fibers, 264
Fibrin, 120
Fibrinogen, 120
Fibrocartilage, 28, 31
Fibroid tumors, 234
Fibrous capsule, 46, 198, 199
Fibrous joints, 46
Fibrous pericardium, 99
Fibrous tissues, 31
Fibula, 64
Filtration, 200-201
Fimbrae, 227
Fissure of the lung, 147
Flat bones, 45
Flatfoot, 90
Flexion, 47
Flexors, 82, 86
Floating ribs, 59, 61
Follicle-stimulating hormone, 226, 242
Fontanel, 51
Food, digestion, 35, 164-191
  disorders, 188
  in large intestine, 183
  in small intestine, 180-182
  in stomach, 173-177
Food transport, 94-132
Foot. See Feet
Foramen ovale, 99
Fracture, bone, 68
Framework, body. See Body framework
Frontal bone, 56
Frontal sinus, 143
FSH, 226
Fundus, 228
Furuncles, 214

Gallbladder, 180
Gallstones, 181, 189
Gametes, 220, 221, 225
Gametogenesis, 220
Gamma globulin, 117
Gangrene, 134, 213
Gastric glands, 175, 176
Gastritis, 188-189
Gastrocnemius muscle, 86
Gastroenteritis, 189
G-Cs, 251
General circulation, 95, 104-106
GH, 242
Gigantism, 258
Gingivae, 168
Glands, 238
  adrenal, 250-251
  digestive, 176
  endocrine, 238-256
  pancreas, 253, 254
  parathyroid, 245, 247
  pituitary, 241-242
  sex, 250, 251
  skin, 207, 208, 209
  thymus, 247
  thyroid, 245-246
Glaucoma, 290
Glenoid fossa, 61
Gliding joints, 47
Globin, 118
Glomerulus, 199
Glucocorticoids, 251
Gluteus maximus muscle, 86, 88

Gluteus medius muscle, 86, 88
Goiter, 257
Golgi apparatus, 12
Gonadotrophic hormone, 229
Gonads, 251
  disorders, 259
Gout (gouty arthritis), 70
Graafian follicles, 225
Granulocyte (granular leukocyte), 119
Gray matter, brain, 269
  spinal cord, 273
Greater omentum, 176
Greenstick fracture, 68
Growth, regulation, 246
Growth hormone, 242

Hammer, 284
Hand, bones, 61, 62
Hard palate, 165
Haversian canals, 52
Head, bones, 56, 58
Hearing function, 284
Heart, 98-101
  disorders, 132-133, 134
Heart block, 133
Heartburn, 188
Heart failure, 133
Heart murmurs, 133-134
Heme, 118
Hemocytoblast, 118
Hemoglobin, 118
Hemophilia, 135
Hemorrhoids, 135
Hepatic portal vein, 106
Hepatitis, 189
Heredity, 11
Hering-Breuer reflex, 152
Hernia, abdominal, 90
Herpes zoster, 214, 287
Herpetic neuralgia, 287
Hiatal hernia, 188
Hilum, 199
Hinge joints, 47
Histiocyte, 119
Hives, 214
Homeostasis, 3
Hormones, 238
  adrenal glands, 250
  pituitary gland, 241, 242
  reproductive system, 225, 226, 229
Human reproduction,
  disorders, 233-235
  organs, 225-230
  reproductive system, 220-223
Humerus, 57, 61
Hyaluronic acid, 222
Hyaluronidase, 222
Hydrocephalus, 288
Hydrochloric acid, 176
Hydrolytic enzymes, 165
Hyoid bone, 166
Hypercalcemia, 247
Hypertension, 135
Hyperthyroidism, 256
Hypertonic solution, 15, 16
Hypophyseal gland, 241
Hypotension, 135
Hypothalamus, 272
Hypothyroidism, 257
Hypotonic solution, 15, 16
Hysterectomy, 234

ICSH, 242
Ileocecal (colic) valve, 183
Ileum, 183
Ilium, 62
Immovable joints, 46
Impetigo contagiosa, 213
Impulse, nerve, 264
Incisors, 169, 170
Infantile paralysis, 288
Infectious causes, respiratory
    disorders, 156-157
Infectious hepatitis, 189
Inferior vena cava, 95, 96 104, 113, 114
Influenza, 157
Infundibulum, 227
Ingestion, 169
Inguinal hernia, 90
Inhalation, 150
Inherited traits, 222, 223
Injuries, bones and joints, 68-71
Innominate bones, 62
Insertion, muscle, 82
Inspiration, 150
Inspiratory reserve volume, 154
Insulin, 238, 253
Integumentary system, 205
Intercellular (interstitial) fluid, 127
Intercostals, 86, 87, 150
Internal orifice (os) of the uterus, 228
Internal respiration, 150
Interpleural space, 147
Interstitial cell-stimulating hormone, 242
Intervertebral discs, 58
Intestines, 196, 197
    large, 183-186
    small, 180-181
Intrinsic factor, 118, 176
Intrinsic muscles, 282
Inunction, 205
Involuntary muscles, 77, 78
Iodine, 245
Iodine tests, 257
Iris, of eye, 283
Iron-deficiency anemia, 134
Irregular bones, 45
Irritability, 264
IRV, 154
Ischium, 62
Islets of Langerhans, 238, 253
Isotonic solution, 16
Isthmus of thyroid gland, 245

Joints, 46
    injuries and diseases, 68-71

Keratin, 206
Kidney failure, 211
Kidneys, 196, 198, 199
Kidney stones, 212
Kyphosis, 70

Lactogenic hormone, 242
Large intestine, 183-186
Laryngitis, 156
Laryngopharynx, 144
Larynx, 143, 145
    cancer, 158
Latissimus dorsi muscle, 86, 87
Laxative, 186
Left cannon carotid artery, 112
Left subclavian artery, 112

Leg bones, 63-64
Lens, of eye, 283
Leukemia, 135
Leukocytes, 52, 118-119
Leukorrhea, 234
Levator, 83
LH, 226, 242
Life function, 2, 3
Ligaments, 47
    injuries, 69
Lipid, 10
Liver, 180-181
Liver bile, 180
Lockjaw, 91
Locomotion, 76
Long bones, 45
Longitudinal arch, 65
Loop of Henle, 199
Lordosis, 70
LTH, 242
Lumbar vertebrae, 58
Lungs, 197
    cancer, 158
Luteinizing hormone, 226, 242
Luteotropin, 242
Lymph, 127
Lymph nodes, 128-129
Lymphatic system, 127-129
Lymphocytes, 129
Lymphoid cells, 129, 247
Lysosome, 12

Male reproductive system, 229-230
Mandible, 58
Manubrium, 59
Marrow, 52
Mastectomy, 233
Maturation, 225
Maxillary sinus, 143
M-Cs, 250
Medial plane, 6
Medulla, 150, 250, 273
Medullary canal, 52
Megakaryocyte, 119
Meiosis, 220
Meissner's corpuscles, 207
Melanin, 207
Melanocyte, 207
Membranes, 10
Menarche, 226
Meninges, 267
Meningitis, 289
Menopause, 226
Menstruation, 226
    disorders, 259
Metabolism, 3
Metacarpal bones, 57, 62
Metarteriole, 113
Metastasis, 233
Metatarsal bones, 57, 64, 65
Microcephalus, 70
Mineralcorticoids, 250
Miner's disease, 158
Mitochondria, 12
Mitosis, 10
Mitral valve, 99
Mixed nerve, 275
Molars, 169
Monocyte, 119
Motion, types of, 47
Motor neurons, 265

Motor nerve, 275
Mouth, role in digestion, 165-166
Movable joints, 46
Mucin, 176
Mucosa, 164
Mucous neck cells, 176
Mucus, 165
Multinucleate, 119
Muscle atrophy, 90
Muscle fatigue, 83, 91
Muscle hypertrophy, 90
Muscle spasms, 91
Muscle tissue, 27, 33
Muscle tone, 83
Muscular dystrophy, 91
Muscular system, 35, 76-89
    abnormal conditions, 90-91
    attachment of muscles, 81-83
    heart muscle, 98
    skeletal muscles, 86-88
Myasthenia gravis, 91
Myeloblast, 119
Myocardial infarction, 133
Myocarditis, 133
Myocardium, 98
Myoglobin, 77
Myometrium, 228
Myxedema, 258

Nasal bone, 56
Nasal cavity, 143
Nasal conchae, 143
Nasal septum, 143
Nasopharynx, 144
Nephric filtrate, 200-201
Nephritis, 212
Nephrons, 199-201
Nerve cells, 18, 264
Nervous system, 264-265
    central, 267-273
    disorders, 287-290
    muscles, contact, 83
    peripheral and autonomic, 275-278
    sense organs — eye and ear, 281-284
Nervous tissue, 27, 33
Neuralgia, 287
Neuritis, 288
Neurons, 264-265
Neutrophil (polymorphonuclear leukocyte), 119
Noninfectious causes, respiratory ailments, 158
Norepinephrine, 251
Nucleolus, 11
Nucleus, cell, 10-11

Occipital bone, 56
Olecranon process, 61
Olfactory nerve, 143
Oogenesis, 220
Orbital cavity, 5, 6,
Orbital socket, 281
Orchitis, 235
Organ of Corti, 284
Organelle, 10
Organs, 2
    body systems, 35-37
Origin, muscle, 82
Oropharynx, 144
Osmolality, 15
Osmosis, 14, 16
Osmotic pressure, 14

Ossa carpi, 61
Ossification, 51
Osteoarthritis, 70
Osteoblast, 247
Osteoclast, 53
Osteocytes, 51, 52
Osteomyelitis, 71
Osteoporosis, 70
Otitis media, 289
Oxyhemoglobin, 118
Ovaries, 238, 239, 250, 251
Ovulation, 226
Ovum, 220, 221
Oxidation, 83, 141
Oxygen, circulatory system, 94-96
    heart, 98-100
Oxygenated blood, 49
Oxytocin, 242

Pacemaker, 100
Pacinian corpuscles, 207
Pancreas, 238, 253-254
    disturbances, 260
Pancreatic juice, 180
Pap (Papanicolaou) smear, 234
Papillae, 166-167
Paralysis, 90
Paranasal sinus, 58
Parasympathetic nervous system, 276
Parathormone, 247
Parathyroid glands, 238, 239, 247
    disturbances, 258
Paresthesia, 288
Parietal bone, 56
Parietal cells, 176
Parotid glands, 168
Partly movable joints, 46, 47
Patella, 64
P.B.I., 257
Pectoralis major muscle, 86, 87
Pelvic cavity, 5
Pelvic girdle, 62-63
Pelvic nerve, 276
Penis, 230
Pepsin, 176, 177
Pepsinogen, 176
Peptic ulcers, 189
Peptone, 177
Pericardial cavity, 4
Pericardial fluid, 99
Pericarditis, 133
Pericardium, 99
Periosteum, 53
Peripheral nervous system, 265, 275-278
Peristalsis, 174
Peritonitis, 190
Peritubular capillaries, 201
Pernicious anemia, 134
Perspiration, 208
Phagocyte, 119
Phagocytosis, 17, 119
Phalanges, 62
Pharyngeal arch, 145
Pharyngitis, 156
Pharynx, 144, 173
Phlebitis, 134
Phosphates, 51
Pia mater, 267
Pigment, 207
Pinocytic vesicle, 12
Pinocytosis, 13, 17

Pituitary gland, 238, 241, 242
    disturbances, 258-259
Pivot joints, 47
Plasma, 116
Pleura, 147
Pleural cavity, 147
Pleural fluid, 147
Pleurisy, 157
Plexuses, 275
Pneumonia, 157
Pneumothorax, 147
Poliomyelitis, 90
Polycythemia, 135
Pons, 273
Portal circulation, 106
Portal vein, 106
Posterior chamber, 293
Posterior lobe, 241
Precapillary sphincter, 113
Progesterone, 225, 251
Projection, 281
Prolactin, 229, 242
Pronation, 47
Prostate gland, 229
Prostatectomy, 235
Prostatitis, 235
Protease, 176
Protein-bound iodine test, 257
Protein manufacture, 11
Proteose, 177
Prothrombin, 117
Proximal convoluted tubule, 199
Pruritus, 213
Psoriasis, 213
Ptyalin, 168
Puberty, 226, 229
Public bone, 62, 63
Pulmonary artery, 104, 108
Pulmonary circulation, 104-105, 108
Pulmonary (visceral) pleura, 147
Pulmonary semilunar valve, 99, 104
Pulmonary vein, 105
Pulmonary ventilation, 104-105
Pulmones, 146
Pulse, 114-115
Pupil, of eye, 283
Purkinje network, 100
Pyelitis, 212
Pyelonephritis, 212
Pyloric sphincter, 174, 177
Pyloric stenosis, 189
Pylorospasm, 174
Pylorus, 174
Pyloric valve, 174

Quadripedal, 58

Radioactive iodine test, 257
Radius, arm bone, 61
Receptors, 278
Rectal columns, 184
Rectus abdominis muscle, 86, 87
Rectus femoris muscle, 86, 87
Red blood cells, 18
Red muscle, 77
Reduction, fracture, 68
Reflex action, 273, 278
Reflex arc, 278
Reflux, 114
Regulators of body functions, 238-261
    adrenal glands and gonads, 250-251

    disorders, 256-260
    endocrine system, 238-239
    pancreas, 253-254
    pituitary glands, 241-242
    thyroid and parathyroid glands, 245-248
Rehabilitation muscles, 90
Renal artery, 199
Renal circulation, 106, 199
Renal column, 199
Renal corpuscle, 199
Renal cortex, 199
Renal fascia, 198
Renal medulla, 199
Renal papilla, 199
Renal pelvis, 199
Renal pyramids, 199
Renal vein, 106
Rennin, 177
Reproduction, 220-235
Reproductive system, 35, 220-223
    disorders, 233-235
    organs, 225-230
Residual volume, air, 154
Respiration, 141-158
Respiratory system, 35, 140-141
    disorders, 156-158
    mechanics of breathing, 151-154
    organs and structure, 143-147
Response, to stimulus, 278
Retina, of eye, 283-284
Retroversion of the uterus, 234
Rh factor, 122
Rhinitis, 158
Ribosome, 11
Rickets, 70
Right lymphatic duct, 128
Ringworm, 213
Rods, of eye, 284
Rotation movement, 47
Rupture, 90
RV, 154

Sacrospinalis muscle, 86, 87
Sacrum, 62
St. Vitus' dance, 287
Salivary glands, 168
Salpingitis, 234
S-A node, 100
Sartorius muscle, 86
Scabies, 213
Scapulae, 59
Sciatica, 288
Sclera, 281-282
Sclerotic coat, 281-282
Scoliosis, 70
Scrotum, 229
Sebaceous glands, acne, 212-213
Sebum, 212
Sedimentation rate, erythrocytes, 123
Selective permeable membrane, 14
Sella turcica, 241
Semicircular canals, 284
Seminal vesicles, 229
Seminiferous tubules, 229
Sense organs — eye and ear, 281-284
Sensory nerve, 275
Sensory neurons, 265
Sensory receptors, 278
Septum, 99
Serosa of the uterus, 228

Serous pericardium, 99
Serratus, muscle, 86, 87
Serum albumin, 117
Serum globulin, 117
Sesamoid bone, 64
Seven-year itch, 213
Sex glands, 238-239, 251
Shingles, 214, 287
Short bones, 45
Shoulder girdle, 59
Sickle cell anemia, 135
Sight, 281-284
Sigmoid column, 184
Silicosis, 158
Simple fracture, 68
Sinoatrial node, 100
Sinuses, 143
Sinusitis, 157
Skeletal (striated) muscle, 76
Skeletal system, 35, 44-47
Skeleton, parts of, 56-65
Skin, 196, 205-209
    disorders, 212-214
Skull, 57-58
Small intestine, digestion in, 180-181
Smell, sense of, 143
Smooth muscles, 77
Soft palate, 165
Solute, 13
Solvent, 14
Somatic cell, 220
Somatotropin, 242
Spastic quadriplegia, 288
Specialized cells, 17-18
Species chromosome number, 220
Sperm, 220, 221-222, 229
Spermatogenesis, 220
Sphenoid bone, 56
Sphenoid sinus, 143
Sphincter muscles, 184
Spina bifida, 70
Spinal cavity, 4
Spinal cord, 273
Spinal nerves, 275
Spine, 59
Spinous process, 59
Spleen, 95-96
Spongy bone, 53
Sprain, 69
Sterility, 234
Sternocleidomastoid muscle, 86, 87
Sternum, 59
Stethoscope, 98
STH, 242
Stiff neck, 91
Stimulus, 278
Stirrup, 284
Stomach, digestion in, 173-177
Stomach serosa, 175-176
Stomatitis, 188
Stratum corneum, 206
Stratum germinatium, 206, 207
Striated muscle cells, 76
Structural units, 1-8
Sty, 290
Subclavian veins, 128
Subluxation, 70
Submucosa, 175
Sudoriferous glands, 208-209
Sugar in urine, tests for, 202
Superior vena cava, 104

Supination, 47
Surfactant, 146
Suspensory ligament, 283
Suture, 56
Sweat glands, 208-209
Sympathetic nervous system, 276-277
Symphysis pubis, 63
Syncytium, 76
Synovial fluid, 46
Systemic circulation, 96, 104-106
Systems, 35-37
Systolic blood pressure, 114, 251

Talipes, 70
Talus, 64
Tarsus, 64
Temporal bone, 56
Teeth, 168-170
Tendons, 47
Testes, 229, 238, 239, 251
    disorders, 235
Testosterone, 229, 251
Tetanus, 91
Tetany, 258
Thalamus, 272
Thoracentesis, 147
Thoracic cavity, 4, 147
Thoracic duct (left lymphatic duct), 128
Thoracic vertebrae, 58
Thrombin, 120
Thrombocytes, 119
Thromboplastin, 120
Thrombosis, 135
Thymus gland, 129
Thyroglobulin, 246
Thyroid gland, 238, 245-246
    disturbances, 256-258
Thyroid-stimulating hormone, 242, 245-246
Thyrotropin, 242, 245-246
Thyroxin, 245, 246
Thyroxin level tests, 257
Tibia, 64
Tibialis anterior muscle, 86, 87
Tidal volume, 154
Tissues, 8, 27-33
Tongue, 166-168
Tonsillitis, 156
Tonsils, 144
Trachea, 145-146
Transportation system, food and oxygen, 94-138
    blood, 94-95, 116-123
    blood vessels, 111-115
    disorders, 132-135
    general circulation, 104-106
    heart, 98-100
    lymphatic system, 127-129
    pulmonary system, 108, 109
Transpyloric plane, 4, 5
Transtubercular plane, 5
Transverse arch, 59, 65
Transverse colon, 184
Transverse processes, 59
Trapezius muscle, 86, 87
Triceps muscle, 86, 87
Tricuspid valve, 99
Triiodothyronine, 245, 246
True rib, 59
TSH (thyroid-stimulating hormone), 242, 245-246

Tuberculosis, 71, 157
    kidneys, 212
Tubules, 199, 200, 201
Tumors, 71
    breast or uterus, 233, 234
Tunica adventitia (external), 111
Tunica intima, 112
Tunica media, 111
Turbinates, 143
Turner's syndrome, 260
TV, 154

Ulna, 61
Universal donor, 121
Universal recipient, 121
Uremia, 211
Ureters, 198, 201
Urethra, 198, 202
Urinary bladder, 202
Urinary meatus, 202
Urinary system, 198-202
Urinary tract, disorders, 211-212
Urticaria, 214
Uterus, 225, 227-229
    disorders, 234
    removal, 234
Uvula, 166

Vacuole, 13
Vagina, 229
Vagus nerve, 276
Valves (in veins), 114
Varicose veins, 134
Vas deferens, 229
Vascular tissues, 32
Vasopressin, 242
Vastus lateralis muscle, 86, 87
Veins, 113-114
Vena cava, 104, 114
Ventral cavity, 4
Ventricle, 99
Venules, 105
Vermiform appendix, 183
Vertebrae, 58, 59
Vertebral column, 58
Villi, 180
Virilism, 259
Viruses, respiratory disorders, 156-157
Viscera, 36
Visceral organs, 276
Vital capacity, breathing, 153-154
Vitamin K, 120-121, 122
Vitreous humor, 283

Waste materials, elimination, 184, 185-186, 196-215
    disorders, 211-214
    excretory system, 196-197
    skin, 205-209
    urinary system, 198-202
Water balance, 272
Webbed fingers and toes, 70
White muscle, 77
White matter, brain, 273
    spinal cord, 273
Wisdom teeth, 169

Xiphoid process, 59, 61

Zona pellucida, 221, 222
Zygote, 222